装备科技译著出版基金

机械手软硬控制策略融合

Fusion of Hard and Soft Control Strategies for the Robotic Hand

[美] Cheng – Hung Chen
Desineni Subbaram Naidu 著

胡劲文 徐钊 赵春晖 译

国防工业出版社

·北京·

著作权合同登记　图字:军 - 2020 - 009 号

图书在版编目(CIP)数据

机械手软硬控制策略融合/(美)陈成雄
(Cheng - Hung Chen),(美)德西内尼·苏巴拉姆·奈都
(Desineni Subbaram Naidu)著;胡劲文,徐钊,赵春
晖译. —北京:国防工业出版社,2023. 2
书名原文:Fusion of Hard and Soft Control
Strategies for the Robotic Hand
ISBN 978 - 7 - 118 - 12771 - 3

Ⅰ. ①机… Ⅱ. ①陈… ②德… ③胡… ④徐… ⑤赵
… Ⅲ. ①工业机械手 Ⅳ. ①TP241. 2

中国国家版本馆 CIP 数据核字(2023)第 029242 号

Fusion of Hard and Soft Control Strategies for the Robotic Hand by Cheng - Hung Chen and Desineni Subbaram Naidu
978 - 1 - 119 - 27359 - 2

※

国防工业出版社出版发行
(北京市海淀区紫竹院南路 23 号　邮政编码 100048)
三河市腾飞印务有限公司印刷
新华书店经售
*
开本 710 × 1000　1/16　**印张** 12¼　**字数** 218 千字
2023 年 2 月第 1 版第 1 次印刷　**印数** 1—1500 册　**定价** 129. 00 元

(本书如有印装错误,我社负责调换)

国防书店:(010)88540777　　书店传真:(010)88540776
发行业务:(010)88540717　　发行传真:(010)88540762

目　录

第1章　简　　介

手被认为是大脑的代理，并且是人体最有趣和最灵活的附属器官。在过去的几十年中，人们通过尝试制作假肢或机械手来代替人手模仿各式各样的类似人的操作，如移动、握紧、抬升、旋转等。完全复制一只人手至今仍是一个巨大的挑战，因为人手结构的复杂性表现在人手由 27 块手骨组成，约由 38 块肌肉控制，拥有 22 个自由度（DOT）；同时，人手还与 4 种约 17000 个触觉单元配合完成其生理作用[1-2]。通过汇集实验心理学家、运动学家、计算机科学家与电气工程及机械工程工程师来研究这些机械手在设计和控制中的传感器运动集成问题，并展现灵巧机械手与人手的相似之处。

在本章中，我们将介绍与军事相关材料，综述控制策略、软硬控制策略的融合，并且概括剩余章节内容。

背景

本书来源于生命科学第三次革命①项目“工程和物理科学与生命科学的融合”（Convergence of Life Sciences，Physical Sciences，and Engineering，CLSPSE）[3-6]。主要内容包括机械手的假肢和非假肢应用[7]。非假肢应用又包括工业、化工上的操作以及核工业危险环境中的应用[8-9]，空间站的建立、修复和维护[10-11]，爆破和恐怖袭击[12]以及机器人手术[13]等。

1.1　军事应用

近几年，在阿富汗和伊拉克战争中，至少 251102 人死亡，532715 人受伤[14]。此外，据美国截肢者联盟（ACA）调查显示[15]：美国大约有 190 万人由于作战行动（如冲突和战争）以及事故或出生缺陷而导致肢体丧失。1996 年，国家卫生统计中心发布的国民健康访问调查（NHIS）研究结果表明[16]：在美国，每 200 人中至少有 1 人接受了截肢手术；在新生儿中，约每 2000 人就有 1 人患有先天的四肢缺陷；每周都有近 3000 人失去肢体；预计到 2050 年年底，

① 第一次革命：分子和细胞生物学。第二次革命：基因组学。

截肢人数将达到 360 万人[17]。

下列文件表明:智能假肢和机械手在军事领域中的需求日益增长。

(1) 在近 60 年中,与过去的截肢手术相比,现代科技成果下的臂或手的截肢手术并未发生明显的改变[18]。美国国防部正积极筹备"假肢研究资助"项目,为研制出由中枢神经系统控制,且在触觉、视觉及表现上都与真实手臂没有差异的人工手臂提供持续的资金支持,进而促进上体假肢研究革命性变革[19]。

(2) 生物革命是美国国防部高级研究计划局(DARPA)所强调的为应对未来紧急事态和国家安全的八大战略重点研究之一[20-21]。特别是"人类辅助神经装置"的研究,可以立刻帮助受伤的老兵来操控他们的假肢。细胞和组织工程是生物革命的一个重要相关领域。

(3) 根据 DARPA 下的国防科学办公室(DFO)计划[22],需要大力开展用于战争伤亡护理并兼顾军民两用的新兴技术的研究,如革命性的假肢、人类辅助神经装置、生物启发多功能动态机器人等技术。另外,文献[23]提到:"现如今,战场上最具破坏性的伤害就是失去肢体……,在 DARPA,未来的版本将变为……重获全功能假肢……"。

2014 年 6 月,在 *IEEE Spectrum* 上发表的一篇文章谈到:"五十年后,我认为我们将在很大程度上消灭残疾。"——伊丽莎·斯特里克兰[24]。机械手除了用于假肢应用外,对于执行真人手不能达到的疲劳阶段的各种操作,特别是对危险废弃材料和产品的处理是非常有用的。

最后,一个关于工程师是如何成为解决方案专家的 IEEE 视频中曾提到:"如果假肢比人体更强壮,更准确,该怎么办呢?"[25]。

1.2 控制策略

1.2.1 假肢/机械手

人工手臂在很多年前就已研制成功,并且被研究人员不断地发展、完善,下列为一些较为完善的假肢/机械手(按照年代顺序)[2,26]。

(1) 俄罗斯手[27-29]。

(2) 早稻田手[30]。

(3) 波士顿手①[31]。

(4) UNB 手(新布伦瑞克大学)[32-34]。

---

① "波士顿手"项目包括哈佛医学院、麻省总医院、自由互助研究康复中心和麻省理工学院。

(5) Hanafusa 手[35]。
(6) 克罗斯利手[36]。
(7) 奥卡达手[37]。
(8) Utah/MIT 手(犹他州大学/麻省理工学院)[38-40]。
(9) JPL/Stanford 手(喷气推进实验室/斯坦福大学)[41-42]。
(10) 明尼苏达州手[43]。
(11) 马努斯手[44-45]。
(12) 小林手[46]。
(13) 罗维塔手[47]。
(14) UT/RAL 手[48]。
(15) 灵巧夹持器[49]。
(16) Belgrade/USC 手(贝尔格莱德大学/南加利福尼亚大学)[50]。
(17) Southampton 手(南安普顿大学,南安普顿,英国)[51]。
(18) MARCUS 手(用户监督下的操作和反应控制)[52]。
(19) Kobe 手(神户大学,日本)[53]。
(20) Robonaut 手(NASA 约翰逊航天中心)[54]。
(21) NTU 手(国立台湾大学)[55]。
(22) 北海道手[56]。
(23) DLR 手(德国航空航天中心)[57-58]。
(24) TUAT/Karlsruhe 手(东京农业科技大学/卡尔斯鲁厄大学)[59]。
(25) BUAA 手(北京航空航天大学)[60]。
(26) TBM 手(麦克米伦儿童康复医院/多伦多)[61]。
(27) ULRG 系统(路易斯安那大学机械钳)[62]。
(28) 牛津手[44]。
(29) IOWA 手(爱荷华大学)[63]。
(30) MA-I 手[64]。
(31) RCH-1(ROBO CASA 手①)[65]。
(32) UB 手(博洛尼亚大学)[66]。
(33) Ottobock SUVA 手(www. ottobock. co. uk)。
(34) 西北大学系统[67]。
(35) 第二代 SKKU 手(成均馆大学,韩国)[68]。
(36) 美国约翰霍普金斯大学(JHU)的应用物理实验室(APL)[23,69-70]。

---

① 关于仿生学机器人研究的意大利-日本联合实验室。

下面是一些关于假肢装置的网站：

(1) Ottobock 的 Sensor Hang™ Speed(www.ottobock.co.uk)。

(2) 隶属于 Otto Bock 集团的 VASI(多能力系统公司)(http://www.vasi.on.ca/index.html)。

(3) 用于运动控制的犹他州手(www.utaharm.com)。

(4) 来自 Touch Bionics 的 i-LIMB Hand(www.touchbionics.com),等等。

文献[2,26]给出了人工手和真人手的对比表,本书对其进行了改编和更新,如表 1.1 所列。

表 1.1　人工手和真人手的比较

(力即抓握力,速度即完全打开和闭环所需时间,E:外部,I:内部)

	尺寸(标准)	手指数量	自由度	传感器数量	执行器数量	重量/gms	力/N	速度/(m/s)	控制
自然人的手	1.0	5	22	≈17000	38(I+E)	≈400	>300	0.25	E
俄罗斯手		5		3	1		147		
早稻田手									
UNB 手									
Hanafusa 手									
克罗斯利手									
Utah/MIT 手	≈2.0	4	16	16	32(E)	—	31.8		E
JPL/Stanford 手	≈1.2	3	9	—	12(E)	1100	45		E
明尼苏达州手									
小林手									
罗维塔手									
UT/RAL 手									
灵巧夹持器									
Belgrade/USC 手	≈1.1	4	4	23+4	4(E)	—	—	—	E
Southampton 手	≈1.0	5	6	—	6(E)	400	38	≈5	E
MARCUS 手	≈1.1	3	2	3	2(I)	—	—	—	I
Kobe 手									
Robonaut 手	≈1.5	5	12+2	43+	14(E)	—	—	—	E
NTU 手	≈1	5	17	35	17(E)	1570	—	—	E

续表

	尺寸（标准）	手指数量	自由度	传感器数量	执行器数量	重量/gms	力/N	速度/（m/s）	控制
北海道手	≈1	5	17	35	17（E）	1570	—	—	E
第二代 DLR 手	>1	5	7	—	7（E）	125	—	—	E
TUAT/Karlsruhe 手	≈1	5	17	—	17（E）	≈120	12	0.1	E
BUAA 手		4			2				
TBM 手									
牛津手									
IOWA 手									
MA - I 手									
RCH - 1 手	≈1	5	16	24	6（E + I）	350	≈40	0.25	E
Ottobock SUVA 手	1	3	1	2	1（E）	600	—	—	I
UB 手									
北海道手	>1	5	7	—	7（E）	125	—	—	E
西北大学系统									
第二代 SKKU 手	1.1	4	4	—	3	900			
APL - JHU 系统									

然而，根据文献[71]，大约35%的截肢患者并不经常使用他们的假肢/机械手。究其原因是多方面的，如现有假肢/机械手功能性不足，或者患者有自卑或恐惧心理等。为了克服这些问题，必须设计出一种无论是外观上还是性能上，都能够尽可能像人类手的假肢。

现已有许多关于肌电假肢/机械手的调查和发表多年的代表最先进技术的文章，包括文献[28]中给出的苏联（俄文）研究成果。此外，文献[2,72 - 84]也给出了一些。

1.2.2 发展历史

文献[85 - 86]对假肢/机械手技术的文献进行了概述，在1.2.3节中进行简要总结。本节着重于最近的发展并不断更新，旨在补充已有的优秀文章[2,79,81,87 - 88]。而且本部分仅为机械手发展的概述，而非详尽的调研，如有遗漏，实属无意。

1970年以前

肌电图（EMG）信号是一种简单且容易获得的关于用于人造/假肢手的各种运动的信息源。相对于植入手术，利用表面电极进行肌电图信号提取的方法对于用户非常有吸引力。针对第二次世界大战中大量伤亡的需要，美国国家科学院开展了在假肢/机械手领域的研究活动[89]。肌电图信号的首次提出是为了控制截肢患者的假肢/机械手[90-91]。开环比例控制系统，即机械手电机的电压及其速度和力从应变片的变化中测量，其中应变片的变化正比于假肢/机械手产生的肌电图信号[92-93]。除此之外，在系统中加入力和速度反馈控制，会让使用者在使用设备时感到更加自然。一种用于南安普顿手(Southampton Hand)的自适应控制模式被开发出来[94]。

1971—1979 年

在文献[32]的研究中，假肢/机械手拇指边缘安装了半导体应变计，通过其提供感知反馈将激励大小调至目标值，并避免在控制假肢过程中出现假肢掉落或者损坏物品等情况。当应变仪接收到刺激时，系统将信号放大并传输到比较器，然后比较器将刺激幅度范围调整到用户需要的水平。然而，带有反馈的设备比正常的手要大两三倍。递阶优化方法[95]包括分析控制理论，例如自适应自组织控制算法和利用模糊自动机技术的人工智能，用来驱动一个机械手。

1980—1989 年

从历史发展的角度来看，文献[72]阐述了支撑假肢/机械手设备应用的闭环(反馈)控制原理的研究状况，讨论了辅助感知反馈、人工反向和控制接口反馈三个方面的相关概念，并得出结论："在过去 65 年里，我们对这些概念的临床应用并未取得显著进展。"文献[96]对肌电信号进行了统计分析，包括零交叉点、二阶到五阶矩、相关函数和模式分类等研究。在整体绝对值(IAV)的特征空间中建立了一个 EMG 模式的概率模型，以提供由运动和速度变量表示的命令与用于假肢/机械手实时控制模式的位置和形状之间的关系，如在文献[97]中。利用运动学关系建立动态模型，采用频域极点配置的多变量反馈控制策略[42]保证 JPL/Stanford hand 的手指稳定性。文献[42]的工作首先建立三指(拇指、食指、中指)和三个关节的动态模型，然后使用拉普拉斯变换在频域工作。为保证控制系统的稳定性，根部必须位于左半平面内。因此，它们可以通过控制根部的位置来获得所需的稳定的手指移动。文献[98-99]报告的工作小组是第一批研究各个方面的小组之一，例如运动学、幻想、动力学和控制多指手三维任意形状的物体。

1990—1999 年

文献[100-101]中提出一种改进的控制臂肘关节的设计、实现和实验验证方法，既包括来自截肢者的内部(自愿的)输入，也包括来自外部环境的输入，来模仿自然肢体。文献[102]考虑设计一只灵巧的手，采用系统方法，通过执行器的位置控制、肌腱张力控制、关节扭矩控制、关节刚度控制，以及笛卡尔指尖刚度控制，来实现刚度控制。文献[75]中对33名使用比例控制肌电手的患者进行了一项调查。根据之前使用终端设备的经验，将其分为三类：数字控制肌电手、身体动力终端设备和无终端设备。调查表明，数字机械手的患者体验组对比例控制机械臂印象最深，因为它有很多优点：舒适、美观，与自动手相比更自然，有更强的按压力度，花费能量更少却功能更多，具有感觉反馈、力反馈，并且肘部以下的部分很短。文献[103]中的研究用三次试验对患者佩戴的神经假手的控制输入输出特性进行了评估：第一次实验测试静态输入输出特性；第二次实验用于测试跟踪步进和斜坡函数时机械手握持的输出控制；第三次实验通过获得手抓系统动力学的输入输出频率响应，以估计转移的频谱分析。当用户控制抓握力并掌握手的位置跟踪时，每一项测试均使用视觉反馈。文献[104]表明：肌电信号(MES)在肌肉收缩的初始阶段不是随机的，因此文献中提供了一种从不同的收缩类型分类模式的方法，即要建立受试者的60个记录，然后产生一些等距收缩类型，如弯曲和扩展。这个信息在设计一个多功能的利用神经网络(ANN)对肌电模式进行分类肌电控制系统时很有用。此外，还研究了隐藏层的大小、段长度和EMG电极的位置。在文献[105-108]中可以发现一些关于使用模式识别方法来进行MES提取和分类的多功能肌电控制系统的工作。文献[50]介绍了一种Belgrade/USC机械手，称之为PRESHAPE(用于人手和假肢评估的可编程机器人实验系统)。该系统利用压力传感器、力传感器和压力反馈将任务命令转换为电机命令，其中压力反馈对小接触力检测非常有用。进一步，文献[109-111]研究了该类多指机械手的控制原理，使其可用于修复与康复。利用非线性神经肌肉(电机伺服)控制系统的动态模型，包括肌肉和拉伸反射的机械性能(如黏滞性)，日本神户大学开发了一种基于表面的肌电信号控制仿生臂(称为“神户手”)，它有三根手指，即拇指、食指和中指，来源于文献[53]中一个由EMG信号处理单元、动态模型、位置控制单元以及修复/机器人装置组成的系统。文献[76]中提及了一个对自主抓取机械手灵巧性、平衡性、稳定性和动态行为四个重要特性的调查。这个调查中很有趣的是一系列与现有的多指机械手、强制闭包、运动学冗余机械手的灵巧性、平衡性、机械控制和稳定性有关的表。文献[112]中一个由动画式假肢发展而来的智能假肢控制系统包括两部分：位于机械臂上的动画控制系统(ACS)和一个远程的能够开启或关闭速度/手柄的假肢充能装置(PCU)。文献[113]中使用一种非线性反馈控制方法来解方程组及线性化，以操纵与滚动接触的对象，完成对两臂的动态控制。

文献[114]中提及了一种由位于意大利博洛尼亚的国家事故研究所开发的基于力感电阻(FSR)的传感控制系统。这个系统是为了控制用于商业的假肢/

机械手对物体的握力。该系统的两个主要功能是:自动搜索与对象的联系;通过非自愿反馈(力传感器和滑动传感器)对物体的检测。文献[115]中使用模糊逻辑(FL)专家系统生成软件包(微处理机控制臂自动调谐)来自动优化控制参数。这个自动调节软件的工作如下:通过客户端连接机械臂硬件,程序用两个传感器信号作为客户端输入,结合存储在 FL 数据库中的定性和定量信息来计算参数值,然后将新的参数值存储到机械臂控制系统存储器中。文献[116]利用一种混合的方法,将离散事件部分与一个变量结构阻抗控制算法相结合,提出了机械手的动态建模。文献[56]提出了一种新颖的在线学习方法,它基于 EMG 测量系统的假肢/机械手控制系统。系统由 3 个单元组成:分析单元,用于生成包含有用信息的特征矢量,并从肌电信号中识别运动;适应单元,用于适应被截肢者的个体变化并从特征矢量中识别运动,同时生成必要的控制命令给假肢/机械手;训练单元,用于指导适应单元根据被截肢者的教学信号和特征矢量实时学习。文献[114]中建立了一个基于 FSR 的感觉控制系统,用于检测运动的上肢假肢和光学传感器。所产生的假肢是"全或无"(开放或闭合)和比例控制类型(力与 EMG 信号之间的关系是线性的)。但是用户必须要注意,对于传统的控制,它使用了自动的(可视的)反馈。这项工作开发了一种无意识的反馈控制,它使用了两种传感器:强度和滑动传感器。如果假肢的手在滑动,控制系统会自动地命令假肢的执行器来增加握力。在接收到 EMG 信号时,手开始闭合动作并继续闭合,直到 FSR 产生一个大于或等于一个接触阈值的信号。然后它停止,因为这个物体已经被需要的合适的握力抓住了。文献[117]的调查表明,被称为"安倍局域网"的神经模糊分类器能够正确识别与人类手部不同动作相关的所有电磁信号。文献[54]中高度拟人化的人类手(Robonaut Hand)由 5 根手指组成,有 14 个独立的自由度。它是在美国国家航空航天局(NASA)约翰逊航天中心建造的,用于与国际空间站(ISS)的外太空活动(EVA)组员接口。

2000—2007 年

在文献[118]中,使用神经网络估计肌肉收缩量和外伸量,开发了一种新的阻抗控制技术[119],来控制阻抗参数,如惯性矩、关节刚度以及一个机械臂的骨骼肌模型的黏滞性。文献[78]概述了机械手的灵巧操作,并绘制了 1960—2000 年间机械手灵巧操作发展的时间线图。文献[77]中发表了一个优秀的调查,总结了机械手的发展和现状,主要关注操作灵巧性、抓握鲁棒性和人类可操作性的功能要求。而且,文献[120]中研究了对灵巧机械手进行滚动运动的构造控制能力算法,和两种具有规则刚性的物体的非完整特性,并对其进行了研究。文献[121-122]提出了一种控制系统架构,其中:基于肌电图测量的前馈环路由低通滤波器和神经网络组成,提供实际转矩信号;基于期望角度的反馈环路由比例-

导数(PD)控制器组成,提供期望的转矩信号。这些力矩之间的误差信号驱动假肢/机械手实现期望角度,而神经网络则根据反馈误差进行学习。文献[123]中研究了手指伸展、外部控制、头顶伸展和前臂内旋。对于手指的伸展,使用了两个电极:一个放在第二掌骨和第三掌骨之间;另一个放在第三掌骨和第四掌骨之间。这样可以使食指、中指和无名指得到充分伸展。对于外部控制,开发了一种新的控制形式,通过保留自主的手腕伸展来控制握力的打开和关闭。头顶伸展是通过刺激肱三头肌来实现的,所以肘部的位置是通过主动激活肱二头肌来控制的。

对于前臂内旋,主要是为了增加通道的数量,以便刺激手指固有肌、肱三头肌和前臂旋前肌,在植入控制源、传感器和身体之间通信的同时,减少体积和裸露线路。文献[2]回顾了临床和研究领域使用肌电信号控制仿生手的传统方法,并指出了假肢控制策略的未来发展,特别是具有生物兼容神经接口的神经假肢,其为用户提供感觉反馈,从而实现基于电子神经图(ENG)的控制取代肌电控制。文献[34]中南安普敦大学与新布里维克大学进行了合作,开发了一种混合控制系统,使用多层感知器 ANN 作为从 MES 提取时域特征集(零交叉、平均绝对值、平均绝对斜率和轨迹长度)的分类器,并使用数字信号处理器(DSP)在没有视觉反馈的情况下控制假肢/机械手的握力。文献[124]中提出设计和开发一种适用于机械臂或机械手的欠驱动机制(驱动器的数量少于自由度),它基于手指的动态模型,使其具有自适应能力(即能够符合手内的物体的形状)。尽管文献[94]已经为南安普顿机械臂(Southampton Hand)开发了一个自适应控制方案,但在之前,他们已经在文献[51]、[79]和[125]的基础上进行了进一步研究,生产出了智能机械手。文献[126]展示了基于微处理器的控制系统的发展过程,分为第一代(基于数字系统)、第二代(低功率)及第三代(基于微处理器和 DSP)。文献[44]对 Oxford 和 Manus 的机械手进行了比较,主要包括:

(1) 手部机械。

(2) 电子控制:模拟放大器、A/D 转换器、DSP 等。

(3) 传感器:基于霍尔效应的力、位置和滑动传感器等。

(4) 操纵或控制方案:Oxford 机械手使用南安普敦的自适应操纵计划,由三层等级制度组成,Manus 机械手使用了一个两级的方案。

文献[127]中建议的方案由五个模块组成,包括人工肌肉骨骼系统、位置和力传感器、3D 力传感器、用于控制滑移和抓取的低层控制环和肌电控制单元。此外,该方案采用两块半导体应变片作为力传感器,将霍尼韦尔国际公司 SS496B 中的传感器作为位置传感器,即线性滑块和小磁铁。此外,控制系统接收三个信号:激活(EDG,用于识别是否有运动)、方向(SGN,决定打开或关闭)和运动的振幅(AMP,作为运动比例控制的输入)。在控制方案上,采用了简单的比例开环控制。

为了保证抓握的稳定性,文献[128]中提出了一种针对圆柱形物体的抓握方式和平行力/位置控制方法。文献[129]提出了一种手肘关节控制假肢手的反馈控制系统。使用扩展生理本体感(EPP)的概念(即使用天然生理传感器),文献[129]和[130]的研究都开发了基于微处理器的上肢假肢控制器。文献[131]对假肢/机械手的研究进行了系统描述,尽管其调查的主体是下肢假肢。此外,文献[128]中设计了一种系统,以获得给定抓握任务的最大负载和接触力为目的,采用平行力/位置控制来保证抓握的稳定性。该控制方案的目标是指定一组关节力矩输入,以实现沿约束方向的期望抓取力和沿无约束方向的期望位置轨迹。

文献[82,132 - 133]中的结果表明,在不知道物体运动学参数和质量中心位置的情况下,利用多手指机械手来实现抓取物体的功能,可以实现动态力/转矩闭合。此外,进一步利用"流形上的稳定性和渐近稳定性"概念,证明了整个手指 - 目标系统的运动收敛性。文献[45]解决了机械设计和操纵(控制)问题,利用欠驱动的运动学增强了机械手性能,并提供了 4 种抓取模式(圆柱、精确、钩和横向),只有两个执行器,一个用于拇指,另一个用于其余的手指。特别是,分层控制体系结构包括一个主控制器,用于 EMG 管理、集点控制(对于位置、扭矩/力)和三个局部控制器,用于低水平关节的刚度控制。在文献[63]中,采用哈林克斯(Haringx)理论和单元刚度模型设计和分析了一种多关节假肢/机械手,三个关节的拇指和两个关节的其他手指,使执行机构的位置可以远离手部,进一步提供了较高的驱动能力和控制自由度。在文献[64]中,加泰罗尼亚理工大学(UPC)的工业与控制工程研究所(IOC)的研究人员设计建造了 MA - I 机械手,其具有 16 个自由度,控制系统由 16 个位置控制回路组成,独立控制 16 个直流电机。基于视觉的手部动作捕捉是一种多维度和多目标搜索优化问题,文献[134]使用了姿势估计和一种运动跟踪方案,通过把遗传算法嵌入粒子滤波(PF)来实现视觉定位手部动作,实现了对机械手的控制。

文献[135 - 136]中提出了一种构想:让不带驱动的机械手(手掌和手指)作为一种软聚合的单一部件,用于自适应性抓取。文献[137]表明,在水平面上平行的物体可以通过一对机械手来控制,从而实现稳定的角度、位置控制和抓取,而不需要像触觉、力、视觉传感器那样的物体参数或对象传感器。在北威斯特大学的机械手实验室(NUPL),文献[138 - 139]中的研究人员开发了多功能假肢/机械手/手臂控制器系统,接收来自多达 16 个植入式肌电传感器(IMES)的信号,并采用启发式 FL 方法[140-141]识别肌电信号模式。特别地,FL 常被用于识别多个表面 EMG 控制信号并将其按用户意图进行分类。多功能机械手包括 3 个手部电机(一个驱动拇指,另一个驱动食指,最后一个驱动中指、无名指和小指),和 2 个手腕电机(一个用于腕关节的运动,另一个用于手腕的转动)。而且,文献[67]表明,在对动力假肢进行 EPP 控制时,减少系统中的静摩擦和侧隙可以防止极限环的出现,其中侧隙的大小由控制索的刚度和动力假肢位于前臂末端的质量决定。

文献[142]证明,通过在周围神经残肢的单个神经束内植入电极,可以刺激被切断的周围神经,向被截肢者提供适当信号的同时,提供有关关节位置和握力的适当的远端参考感觉反馈。文献[143]对具有类人操作能力的仿生手的机械结构、设计和控制系统进行了研究,并对自然手和假肢/机械手进行了比较(详见文献[144-145])。在文献[146]中,利用参数自回归(AR)模型和基于 Levenberg-Marquardt(LM)的神经网络开发了肌电信号运动模式分类器,以识别拇指、食指和中指的三种运动类型,进而控制未驱动的假肢/机械手。

文献[147]主要讨论了控制的"最佳"延迟,即从命令到手运动的最大时间量,对于用户所接受的时间范围为 200~400ms。另外,为了使健全的受试者能够操作假肢/机器人终端设备,有人还开发了一种名为"健全受试者假肢手"(PHABS)的旁路假体。该控制器是市售的 Myo-pulse 控制器,它结合了脉冲宽度调制(PWM)和脉冲周期调制(PPM),提供了速度与数字控制信号的脉冲宽度和时序之间的线性关系,另外还使用了机械低通滤波器来平滑脉冲序列和运动,如果肌电图达到阈值,则电机将"打开"。此外,还在 MATLAB 的 Simulink 中创建了实验控制器,并使用 Simulink Real Time 和 XPC Target Toolboxes 来执行。最后,本书总结了 7 个时间延迟来源,具体如下:

(1) 从运动信号发出到肌电图响应的时间;

(2) EMG 前置放大器中模拟滤波器的时间常数;

(3) 模数采样周期;

(4) 收集用于特征提取的 EMG 信号所需的时间;

(5) 对 EMG 信号进行特征提取所需的时间;

(6) 执行提取特征图案识别所需的时间;

(7) 启动组件所需的时间。

在文献[81]中,对传统的控制方法以及新的控制技术现状进行了回顾。文献[148]开发了一种新的智能柔性手部系统,该系统有 3 根手指,10 个关节,配备有小谐波齿轮传动装置和高功率微型执行器,提供了 12 个自由度,用于抓握任务。文献[149]中开发了一种基于肌电图(使用电极、扭矩和角度传感器)的假肢/机械手控制系统,由操作员、五指驱动假肢/机械手系统、假肢/机械手控制器(包括模数转换器、DSP 板和步进电机)以及视觉反馈组成,其中,研究人员利用带有参数自回归模型和小波变换的神经网络对肌电信号进行了特征提取和特征分类。进一步地,文献[150]提出了一种具有高级监督控制器的分层控制系统,主要用于实现肌电信号采集和模式识别,并向低级控制器提供一组命令(用于操作,如关闭、打开等)。文献[151]提出了一种基于传感器的混合控制策略(使用基于传感器的肌电信号和反馈给用户的正常反馈控制),其中由假肢信号控制的数字控制器将用户抓取意图(肌电信号)转换为控制假肢的顺序。

文献[68]研究开发了带有触觉传感器(滑动传感器和力传感器)的机械手,称为第二代 SKKU 手,它具有两个功能单元:一个是基于聚偏氟乙烯(PVDF)的滑动传感器,主要用于检测滑动;另一个是薄型柔性力传感器,其使用压力可变电阻油墨来读取接触力的大小和物体的几何信息。文献[152]研究了用生物电子学的方法来设计和控制拟人手关闭手指抓住一个对象的情况,利用了参考轨迹和两种不同版本的 PD 控制系统(关节空间和滑块空间)。特别地,机械手由 3 根欠驱动的手指(食指、中指和拇指)组成,这 3 根手指由位于手臂下部的 3 个电缆驱动的直流电机驱动。文献[153]的工作研究了由多功能假体控制器产生的大的控制器延迟。使用一种叫作 PHABS 的装置来测试 20 名健全受试者在盒块试验①中的表现,为了估计和比较假肢/机械手的性能,文献[147]提出了功能指数。文献[154]研究了一种水下柔性机械手(HEU 手 2 号)操作,其利用基于位置的神经网络阻抗控制(PBNNIC)进行力跟踪控制。

在文献[154]中研究人员开发出了灵巧的水下机械手。其传感器系统主要包括 12 个不同位置的应变计,当机械手处于水下时,由于无法准确掌握其完整的动力学模型,控制系统更加复杂。因此,控制系统考虑了机器人动态模型的不确定性。手动跟踪控制器采用 PBNNIC 方案设计。文献[155]利用生物学启发原理设计和控制仿生机械臂,提出了基于机器人动力学的轨迹规划和优化方法。

另一种学习控制策略是在文献[156]中提出的,该策略基于一个假设,即人类运动指令和感觉信息都以离散的间断的方式传递,并通过一种名为 S - learning 的基于序列的学习算法及时量化,这与传统的控制方法不同,因为在实际中机器人的动力学模型是高度非线性的,并且有众多的自由度(DOF)。在文献[157]中,使用一种新的"弦和弹簧"机制和一个连续的小波变换(CWT)来描述

① 医学上用来测试患者手的灵巧度的方法。——译者注。

第一个只装配 3 个电动机但实现了 20 个自由度的五指假肢/机械手模型，其中 EMG 输入用于前馈，反向传播 NN 来识别抓握的类型。

文献[158]的工作重点在于手的控制系统和手的设计优化，提出机械手动作的控制量与电极提取的表面肌电信号强度成正比的假设，其信号提取方法为，将表面电极施加到残肢的一组拮抗肌上。此文献首先解释了手动原型的设计，包括生物机电设计方法、欠驱动人工手部、3D CAD 模型（使用 Pro-Engineer 软件）和动力学分析（使用 ANSYS 软件）。其次，建立了 PD 控制系统模型，包括关节空间和带弹性补偿的 PD 控制。第三，在多目标问题（四目标）下验证和优化手部设计。前两个目标与闭环控制性能相关，其余两个目标是联合轨迹的一部分。另外还在 MATLAB/Simulink 中进行仿真开发。最后，将实验结果与仿真进行了比较。

文献[159]介绍了一种考虑末端质量的非线性柔性机械臂的动态系统，并设计了智能最优控制器，其利用模糊神经网络控制器和鲁棒控制器分别学习非线性函数并补偿逼近误差，可以很好地控制周期运动弯曲振动和扭转振动的耦合。为了克服传统算法规则库大、训练时间长等困难，文献[160]提出了一种具有动态等式约束的自学习动态模糊网络（DFN），加快了智能非线性最优控制的轨迹计算。对于带有肌腱传输的五指欠驱动假肢机械手，文献[161]提出了一个强大的控制器，实现了两个不同阶段的任务，包括手的预成型和涉及手臂快速抓取物体的功能。

1.2.3 2007 年以来主要控制技术综述

1.2.3.1 硬计算策略

1）PD 控制器

Rong 等[162]提出了一种基于自适应前馈控制的 PD 控制器用于控制不确定参数的二自由度直接驱动机器人。

2）自适应控制器

Cai 等[163]针对不可测速度和不确定环境下的双关节机械臂，提出了一种观测器反推自适应控制方案，并基于状态反馈控制器设计了自适应速度观测器，以补偿误差估计。Seo 和 Akella[164]推导出了一种新的自适应控制解决方案，它涉及一种新的多自由度（n - DOF）机器人操纵器系统回归矩阵设计。Liuzoo 和 Tomei[165]针对不确定动态二自由度机器人，通过对每个关节的输入参考信号进行傅里叶级数展开，设计了一种具有不确定动力学的二自由度平面机器人全局输出误差反馈自适应学习控制方法。为实现跟踪的目的，Chenet 等[166]提出了一种具有系统不确定性和扰动的轮式移动机器人的自适应滑模动态控制器，使得轮式移动机器人的实际速度达到期望的速度。

3）鲁棒控制器

由于机械臂关节的粘弹特性，Torabi 和 Jahed[167]采用环整形方法，降低了单关节机械臂鲁棒控制模型在时域和频域上的阶数。为了增强对假肢/机械手的控制，Engeberg 和 Meek[168-171]提出了鲁棒滑模、后退和混合滑模 - 后退(HSMBS)平行力 - 速度控制器，使人能够更容易地控制物体。此外，Ziaei 等[172]开发了建模方法，采用广义正交基函数(GOBF)进行系统辨识，并为单个柔性链式(SFL)机械手开发了鲁棒的位置和力控制器。Jiang 和 Ge[173]通过近似线性化算法，将具有不确定扰动的三自由度移动机器人的非线性运动学模型转化为线性控制系统，然后通过线性矩阵不等式(LMI)设计了部分反馈鲁棒控制器。

4）最优控制器

Vitiello 等[174]结合位置控制器和卡尔曼滤波器，使得 NEURARM 机器人可以进行平面运动，如移动到指定位置和接球。NEURARM 机器人通过连接在电缆上的非线性弹簧进行液压活塞驱动。Vrabie 等[175]受到生物学的启发，提出了一种参与者/批判者结构的非线性方法，在系统内部动力学知识未知的情况下，通过求解 Riccati 方程，解决了连续时间自适应最优控制问题。为了最小化欠驱动二自由度机器臂在输入约束和结构参数约束下的定位时间(在两个特定点间移动)，Cruz Villar 等[176]提出了一种结合结构化参数和 bang - bang 控制率的并行结构控制再设计的方法。针对六自由度电驱动并联机械臂，Duchaine 等[177]通过一般的预测控制律导出了机器人的位置跟踪和速度控制算法、机器人的动力学模型、预测和控制范围以及约束条件，并通过基于计算效率的模型预测控制方案，最终导出了六自由度电缆驱动并联机器人最优控制的解析解。

5）分层控制器

Fainekos 等[178]提出了分层控制率，使用二阶动力学方法跟踪一个具有全局有界误差的单纯动力学模型，并用于求解移动机器人的运动规划问题。此外，还利用自动机理论和简单局部矢量场方法，解决了运动学模型的鲁棒时序逻辑路径规划问题。

1.2.3.2 软计算策略

1）模糊逻辑

根据人体解剖学，Arslan 等[179]建立了三自由度人类食指肌腱结构的生物学模型和模糊滑模控制器，通过调整模糊逻辑单元的滑动面斜率来产生闭合和开放运动时所需的动力。

2）人工神经网络

Onozato 和 Maeda[180]利用两个神经网络，学习用逆动力学来控制二自由度 SCARA 机器人的位置移动。Aggarwal 等[181]对猕猴的每根手指的屈伸、手腕的旋转和灵活抓握这三个不同运动的神经元进行了记录，然后进行了实时的 MATLAB/Simulink 仿真，运用多层前馈人工神经网络，为每一个动作都设计了单独的解码滤波器。Tan 等[182]提出了一种不依赖动力学模型的方法，用于大型不确定性机器人操作系统的在线分散神经网络控制设计。Kato 等[183]阐述了大脑对自适应性假肢/机器人系统的操作方式，通过人工神经网络训练，对肌电信号进行模式识别，用于驱动十三自由度的假肢/机械手。此外，利用功能性磁共振成像（f - MRI）分析人脑与假肢/机械手之间的相互适应关系，根据上肢的变化，分析得出大脑感知皮层区域的可塑性。

3）遗传算法

Marcos 等[184]提出了基于遗传算法的闭环逆向法（CLGA），用于最小化三自由度冗余机器人轨迹规划中两个相邻构型的最大关节位移、关节速度、关节加速度、总关节力矩和总联合功耗。此外，Kamikawa 和 Maeno[185]运用遗传算法优化枢轴位置和握力，并为欠驱动机械臂设计了一台超声波电机来控制 15 个柔性关节。

4）粒子群优化算法

Khushaba 等[186]提出了一个基于粒子群优化算法的肌电控制假肢/机械手系统。通常，由于机械臂的抓取精度有限，例如很难抓住螺钉或针，因此需要对指尖轨迹和控制系统的精度和有效性进行优化。

1.2.3.3 软计算和硬计算融合策略

1）比例 - 积分 - 微分（PID）控制器和鲁棒控制器

Dieulot 和 Colas[187]开展了一个针对柔性轴的鲁棒参数方法设计的案例研究，设计了基于刚性模式附加极点配置约束的 PID 控制器的启发式初始整定方法。

2）自适应控制器与鲁棒控制器

为了实现在未知摩擦和不确定性影响下的轨迹跟踪任务，Chen 等[188]采用了一种复合跟踪方案，通过自适应摩擦估计来确定库仑摩擦、黏性摩擦和斯特里贝克（Stribeck）效应，并使用鲁棒控制器来增强二自由度平面机器人机械手的整体稳定性和鲁棒性。

3）鲁棒控制器与最优控制器

Huang 等[189]设计了具有一定不确定性的鲁棒控制系统，包括未知的物体载荷、模型和动态参数，并将其作为空间机器人捕捉未知物体的最优控制方法优化的性能指标。

4）鲁棒控制器与模糊逻辑

Tootoonchi 等[190]将鲁棒定量反馈理论（QFT）与模糊逻辑控制器（FLC）相结合，以跟踪所需的轨迹，降低了二自由度机械臂系统动力学的复杂性。Yagiz 和 Hacioglu[191]提出了三自由度空间机器人滑模控制器的控制增益自适应方法，增益的大小可以由模糊控制器和滑动面根据系统的误差进行调整，其中滑动面的斜率由 FL 算法动态改变。

5）鲁棒控制器和人工神经网络

Siqueira 和 Terra[192]开发了一种基于神经网络的 H_∞ 控制器，该控制器在近似了实际欠驱动协同操纵器的不确定因素，并在保持一个关节不动的条件下，实现了对操纵器和物体之间位置误差和挤压力的鲁棒控制。

6）滑模控制器与遗传算法

Chen 和 Chang[193]利用多重交叉遗传算法来估计未知的系统参数，并利用滑模控制方法来克服双关节机器人控制的不确定性问题。

7）滑模控制器和粒子群优化

Salehi 等[194]利用在线粒子群优化（PSO）方法，对未知环境下二自由度平面机械臂末端执行器力矩的滑模控制参数进行优化。

8）模糊逻辑和人工神经网络

Subudhi 和 Morris[195]提出了一种混合模糊神经网络控制（HFNC）的方案，这种方案使用一个模糊逻辑控制器和一个神经网络控制器来平衡多关节柔性机械臂出现的刚柔耦合效应。

9）人工神经网络和粒子群算法

Wen 等[196]运用混合粒子群优化算法的神经网络（HPSONN），计算平面二自由度机械手逆运动学控制的伪逆雅可比矩阵。

1.2.4 革命性的假肢

2009 年（见文献[23,69]），位于马里兰州巴尔的摩市的约翰霍普金斯大学（JHU）的应用物理实验室（APL）收到了来自 DARPA（美国国防部高级研究计划局）为 2009 年革命性假肢计划提供的资助。美国国防部的目的是“开发一套新型的能够模仿和感知真实事物的机械臂”。应用物理实验室领导了一个大约由 30 个组织构成的国际团队，其成员主要来自奥地利、加拿大、德国、意大利、瑞典和美国。应用物理实验室的团队交付了第一代 DARPA Limb Proto 1（见文献[70]）。它是一套完整的肢体系统，其中还包括一个虚拟环境，用于病人培训、临床配置，并在临床研究中记录肢体运动和控制信号。

1.3 智能控制策略的融合

这里给出目前对于融合控制策略的智能假肢/机械手的研究成果,其工作原理图如图 1.1 所示(参见文献[34,149,151]的工作)。简而言之,整个系统总体上包括:首先从用户手臂获取肌电图信号,用于表皮电极或植入电极(在植入的情况下,专注于基于纳米材料的生物相容性研究);然后,对肌电信号进行特征提取和分类或者识别假肢/机械手进行不同动作对应的肌电信号;接着,使用识别出的分类信号来控制假肢/机械手的执行器和驱动机构。值得一提的是,对肌电信号的提取、识别和控制算法的研究,利用了软计算(SC)和硬计算/控制(HC)策略的融合。

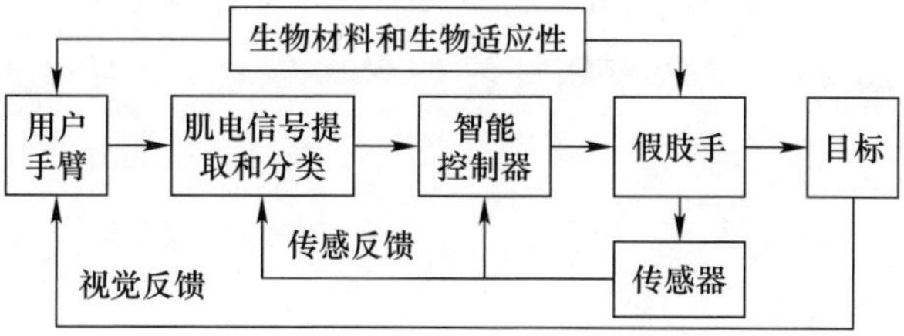

图 1.1 假肢/机器臂技术示意图

1.3.1 硬计算与软计算的融合/控制策略

软硬计算融合策略用于较低级别的精度、准确度、稳定性和鲁棒性的控制,包括 PD 控制[197]、PID 控制[198-199]、最优控制[199-202]、自适应控制[203-206]等,这种策略特别适用于机械臂装置。作者对机器人和假肢/机械手的控制策略进行了综述[85-86]。然而,之前对于机械手的研究工作[197-199,207]表明,使用 PID 控制器会导致超调和振荡等不可取的特性出现。Subudhi 和 Morris[195]对于双关节柔性机器人机械手的研究,以及 Liu 和 Chen[208]对于一个六自由度水下机器人(自动潜航器)的研究也都证明了这一点。

软计算一词或计算智能(CI)在 1994 年第一次被 L. A. Zadeh 使用,他对软计算的定义为,“软计算是一种旨在利用对不精确、不确定、部分事实的容错,去实现可追踪性、近似鲁棒性、低成本的解决方案和更贴近现实的方法的集合”[209]。软计算的基本概念已经受到 Zadeh 早期发表刊物[210-212]的影响。自 1994 以来,许多研究人员和工程师对软计算进行了不同的研究。

与硬计算不同,软计算策略的目的是适应不精确、不确定、只知道部分事实的近似环境[209]。在 L. Magdalena 的评论文章中，分析、比较并讨论了一些文献中对软计算的定义[213]。与硬计算的低级别的控制不同，软计算用于需要人类参与和决策的整体任务的高层次控制。软计算是一个基于协同并集成了神经网络、模糊逻辑和优化方法（如遗传算法和粒子群算法[197,209,213-220]）的新兴领域。之前关于假肢/机械手的研究工作，主要利用神经网络[33-34,221]、模糊逻辑[140-141,222]和遗传算法[223]等，用于机械手各种运动或功能的肌电信号分类。

通过类比大脑与硬计算和软计算策略融合的对应关系，我们提出了[215-216]软计算策略的上层控制和传统硬计算策略的下层控制混合结构的混合智能控制策略。软计算和硬计算方法的融合可以解决单独使用硬计算或软计算方法无法得到满意解决的问题，并实现高性能、鲁棒性、自主性和低成本的要求，如指尖轨迹和控制系统的准确性和有效性[215,216]。对于机械手混合智能控制策略，也可以应用于危险环境、外科手术和临床假肢/机械手等机器人技术中[224-226]。

图 1.2 所示为硬计算和软计算的融合策略，具有以下令人满意的特性[215-216]。

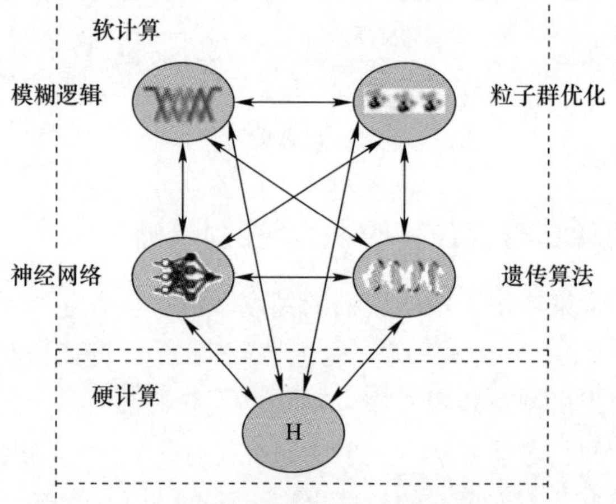

图 1.2　软计算与硬控制策略的融合

（1）基于软计算的方法，特别是模糊逻辑，用于需要人类参与和决策的整体任务的高层次控制上，而硬计算用于准确性、精确性、稳定性和鲁棒性的较低级别的控制。

（2）在使用混合方案的情况下，利用含有神经网络的软计算可以对导弹自动驾驶仪的线性、可变增益控制器提供的控制进行补充。

（3）运用基于软计算的遗传算法来调整 PID 控制器的参数，并在宽波段条件下实现良好的性能和鲁棒性。

(4) 软计算和硬计算是潜在的互补方法。

(5) 这种融合可以解决单独使用哪种方法都不能满意解决的问题。

(6) 新型软计算和硬计算的协同组合能够完成高性能、鲁棒性、自主性和低成本的任务。

我们的研究重点是通过神经网络、模糊逻辑和遗传算法的软计算和硬计算策略的融合(见文献[216 - 227]),开发用于肌电信号提取、分析和假肢控制的智能自主策略。大体上而言,该方法充分利用了我们在假肢方面的研究经验,如文献[228 - 229]所述,特别是如文献[230 - 231]所述的生物医学工程中的问题。

文献[232]综述了9篇在工业和工程应用中使用这些策略的论文。对于融合策略,从5个方面阐述了文献[233]中描述的多维度分类方案:软计算和硬计算组件的互联程度(融合级别)、融合技术的拓扑结构(融合结构)、融合发生的时间(融合时间)、系统架构的层级(融合级别)以及应用程序的动机(融合激励)。此外,还将融合策略分为12大类和6大补充范畴[234]。

1.4 我们所做的研究工作

我们所做的研究工作按时间顺序概述如下:

Lai 等[235]在一次简短的回忆中提到,由机器人植入的生物界面和其他假肢/机器人设备的重要性,并提出一个跨学科的由生物医学、组织工程、生物材料和生物医学领域的科学家组成的团队,需要在一起协同工作,才能取得成效。

针对粒子群优化算法,我们研究了一组基于粒子群优化算法的算子,以寻找一些经典基准问题的最优值。粒子群算法作为数学方法的实现,受到了具有社会行为的鸟群和鱼群的启发。此外,粒子群算法利用少量且不那么复杂的规则来对需要大量内存和处理时间的复杂行为进行响应。在粒子群优化算法中,粒子在连续变量空间上,每个粒子都与它的邻近粒子连接,所以粒子的更新速度与仿真结果密切相关。经过统计分析得到了一个在连续变量空间中粒子群优化算法的速度更新公式。特别是,我们研究了影响粒子更新速度的概率密度函数以及用于构建群内每个粒子的更新速度矢量的分量。几个数值算例的仿真结果表明,在全局最优条件下,必须获得少量的负速度才能获得最佳的最优值[219]。

我们对控制理论在假肢/机械手上的应用进行了综述:如多变量反馈,最优、非线性、自适应、鲁棒性的硬计算策略,以及如人工智能、神经网络、模糊逻辑、遗传算法、粒子群算法的软计算策略;还有硬软控制策略的融合[85]。文献[197]提出了一种粒子群优化算法,用于计算白细胞黏附分子破裂力,同时又提出如何确定机械手正确控制参数这一问题。另外一项研究是,由伊利诺伊州立大学工作组提出的运用在假肢/机械手上的遗传算法和 PID 控制算法协同的软计算和硬计算融合策略。值得一提的是,一种自适应神经模糊推理系统(ANFIS)被运用于三关节食指的运动学控制,用于手部运动的反馈线性化算法,以及求解 PID 控制器最优参数的遗传算法[198]。基于每一次迭代改变最大速度的一种自适应粒子群优化(APSO)算法,用于求解两个 30 维度的基准问题[220]。

这项研究展示了一种使用食指和拇指进行二维运动,安装在假肢/机械手上的混合自适应神经模糊推理系统和自适应控制的软计算和硬计算融合策略。推导了假肢/机械手的动力学模型,并利用反馈线性化技术获得了线性跟踪的动态误差。然后设计出一种自适应控制器以减小跟踪误差。与 PID 控制器相比,这种混合控制器的性能有了明显的提高。自适应控制策略能够扩展应用在具有十四自由度且质量和惯性都未知的振动智能假肢/机械手上。仿真结果表明,装有自适应控制器的假肢/机械手可以准确地抓握对象,并且无超调振荡[236]。

一种融合增强型连续禁忌搜索(ECTS)算法和粒子群优化算法的浓缩型混合优化(CHO)算法被研究出来[207]。该浓缩型混合优化算法结合了增强型连续禁忌搜索算法和粒子群优化算法的各自优势。需要强调的是,首先使用增强型连续禁忌搜索算法来确定较小的搜索空间,接着运用基本的粒子群优化算法来寻求各自的局部最优解。增强型连续禁忌搜索算法采用多样化的禁忌搜索概念来遍历全局搜索空间,然后在搜索空间中选取有用的域。一旦确定了搜索空间中有用的域,另一种浓缩型混合优化算法使用强化概念的禁忌搜索,以最终确定期望的域。这种浓缩型混合优化算法通过了多维度双曲面的 Rosenbrock 问题的测试。与其他四种算法相比,该算法具有更好的准确性和有效性。另一种融合自适应神经模糊推理系统软计算和硬计算技术的算法,采用了有限时间内线性二次最优控制对假肢/机械手的双指(拇指和食指)进行研究[201,237-238]。特别地,自适应神经模糊推理系统用于逆运动学,并使用反馈线性化动态最优控制,以尽量减少跟踪误差。与 PID 控制器相比,这种混合控制器的仿真表现出了更好的性能。这项工作被扩展到一个五指、三维的假肢/机械手上[199]。为了使最优控制器快速动作并提高精度,将包含指数项[202,239]的性能指标 J 进行修正。仿真结果表明,所提出的方法得到了快速响应,具有较高的精度并有比自适应神经模糊推理系统或基于遗传算法计算速度快 30 倍的轨迹规划[201,239-240]。

1.5 神经义肢的成果

值得注意的是,在文献[241-244]中介绍了一些神经假肢的发展。

一个有趣的研究是文献[245],它采用功能性电神经假体(FES)植入,为四肢瘫痪的人提供抓取和放松功能,并且比较了肩膀、手腕、肌电腕伸肌位置的三种不同的控制方法。为了提高 C5 和 C6 部位脊髓损伤的人在把握强度、前臂旋转和肘部屈伸方面的控制作用,文献[123]研究开发出了一套先进的神经义肢。它包括植入的零件、十通道的激励器、引线和电极、一个关节角传感器,以及如控制单元和收发线圈之类的外部元件。

值得一提的是,在文献[246-248]中提到,Jesse Sullivan 在一次电气事故中失去了双臂,他通过切断大脑中控制手臂和手掌移动的基本神经,重新把控制作用定位于胸腔肌肉神经上,成功地移动了他的仿生手臂。连接在胸部肌肉上的电极产生了一个电信号,这个电信号的作用是根据肌肉运动性质控制仿生手臂,并且它的特性是在大脑中"思考"要用手臂做什么。然而,所展示的仿生手臂只是一个"原型",只能用于研究。

另一个有趣的新闻出现在文献[249-250]中,它报道了在一个四肢瘫痪的人的大脑中植入电子芯片,并使用计算机来操纵仿生手臂。

2014 年 9 月,*IEEE Spectrum*[251]中刊登了一篇文章,描述了"一个癫痫病患者通过脑电波控制机械肢体"。

1.6 总　　结

本书由 7 章组成。第 2 章介绍运动学与轨迹规划;第 3 章给出机械手的动态模型;第 4 章讨论模糊逻辑、神经网络、自适应神经模糊推理系统、遗传算法和粒子群优化算法等软计算策略;第 5 章和第 6 章介绍机械手和每根手指的软和硬控制策略融合。最后,在第 7 章给出结论和对未来工作的一些想法。

参考文献

[1] E. R. Kandel and J. H. Schartz. *Principles of Neural Science, Third Edition*. Elsevier/North - Holland, New York, USA, 1985.

[2] M. Zecca, S. Micera, M. C. Carrozza, and P. Dario. Control of multifunctional prosthetic hands by processing the electromyographic signal. *Critical Reviews™ in Biomedical Engineering*, 30:459-485, 2002 (Review article with 96 references).

[3] MIT. *The Third Revolution: The Convergence of Life Sciences, Physical Sciences and Engineering.* Massachusetts Institute of Technology (MIT), Cambridge, Massachusetts, USA, 2011 (The First Revolution: Molecular and Cellular Biology and the Second Revolution: Genomics).

[4] College of Fellows American Institute for Medical and Biological Engineering. Medical and biological engineering in the next 20 years: The promise and the challenges. *IEEE Transactions on Biomedical Engineering*, 60(7): 1767 – 1775, July 2013.

[5] B. He, R. Baird, R. Butera, A. Datta, S. George, B. Hecht et al. Grand challenges in interfacing engineering with life sciences and medicine. *IEEE Transactions on Biomedical Engineering*, 60 (3): 589 – 598, March 2013.

[6] Committee on Key Challenge Areas for Convergence and Health. *Convergence: Facilitating Transdisciplinary Integration of Life Sciences, Physical Sciences, Engineering, and Beyond.* National Research Council (NRC) of the National Academies, Washington, District of Columbia, USA, 2014.

[7] What are the applications of an industrial robot arm? http://www. robots. com.

[8] M. Jamshidi and P. J. Eicker, editors. *Robotics and remote systems for hazardous environments.* Prentice Hall PTR, Upper Saddle River, New Jersey, USA, 1993.

[9] B. Brumson. Chemical and hazardous material handling robotics. Technical note, Robotic Industries Association, Ann Arbor, Michigan, USA, January 18, 2007 (http://www. robotics. org).

[10] S. Bouchard. Robotic arms help upgrade international space station. Technical note, IEEE Spectrum, New York, USA, July 22, 2009 (http://spectrum. ieee. org/automaton/robotics/humanoids).

[11] E. Guizzo. How robonaut 2 will help astronauts in space. Technical note, IEEE Spectrum, New York, USA, February 23, 2011 (http://spectrum. ieee. org/automaton/robotics/humanoids).

[12] Homeland Security Newswire, Mineola, New York, USA. *Homeland Security Newswire – Robotics*, 2015 (http://www. homelandsecuritynewswire. com/topics/robotics).

[13] U. S. Food and Drug Administration, Silver Spring, Maryland, USA. *Computer – Assisted Surgical Systems: Common uses of Robotically – Assisted Surgical (RAS) Devices*, March 11, 2015.

[14] Casualties in Afghanistan and Iraq. www. unknownnews. net, June 5, 2006.

[15] Amputee Coalition of America (ACA) National Limb Loss Information Center (NLLIC) Limb Loss Facts in the United States. http://www. amputeecoalition. Org, March 10, 2011.

[16] P. F. Adams, G. E. Hendershot, and M. A. Marano. Current estimates from the national health interview survey, 1996 *National Center for Health Statistics. Vital and Health Statistics*, 200(10): 1 – 203, 1999.

[17] K. Ziegler – Graham, E. J. MacKenzie, P. L. Ephraim, T. G. Travison, and R. Brookmeyer. Estimating the prevalence of limb loss in the United States: 2005 to 2050. *Archives of Physical Medicine and Rehabilitation*, 89: 422 – 429, March 2008.

[18] D. Moniz. Arm Amputees Rely on Old Devices. USA Today: 10 – 06 – 2005, June 2005.

[19] D. Moniz. Military to Fund Prosthetics Research. USA Today: 10 – 06 – 2005, June 2005.

[20] Director of Defense Advanced Research Projects Agency (DARPA) Presentation to Subcommittee on Terrorism, Unconventional Threats and Capabilities, House Armed Services Committee, United States House of Representatives – Bridging the Gap. Press Release: Dated March 25, 2004.

[21] Director of Defense Advanced Research Projects Agency (DARPA): Bridging The Gap Powered by Ideas. Press Release: Dated February 2005.

[22] Director of Defense Advanced Research Projects Agency (DARPA): Defense Sciences Office, 2006.
[23] DARPA – News Release: DARPA Initiates Revolutionary Prosthetic Programs. Press Release: Dated February 8, 2006.
[24] E. Strickland. The end of disability: Prosthetics and neural interfaces will do away with biologys failings. *IEEE Spectrum*, 6:30 – 35, June 2014. (The Next Fifty Years: THE FUTURE WE DESERVE – A Special Report – BIOMED).
[25] IEEE, New York, USA. *IEEE Solutionists Showing What Engineers Do*, July 21, 2011. (https://ieeetv.ieee.org/ieeetv – specials/ieee – solutionists).
[26] M. Zecca, S. Roccella, G. Cappiello, K. Ito, K. Imanishi, H. Miwa, C. Carrozza, P. Dario, and A. Takanishi. From the human hand to a humanoid hand: Biologically inspired approach for the development of robocasa hand #1. Technical report, 3ARTS Lab, Scuola Superiore Sant Anna, Pisa, Italy, 2006.
[27] A. E. Kobrinski. "Problems of bioelectric control," in *Proceedings of First IFAC*, p. 619, Moscow, USSR, 1960.
[28] E. D. Sherman. A Russian bioelectric – controlled prosthesis. *Canadian Medical Association Journal (CMAJ)*, 91:1268 – 1270, December 12, 1964.
[29] D. S. McKenzie. The Russian myo – electric arm. *The Journal of Bone and Joint Surgery*, 47:418 – 420, August 1965.
[30] I. Kato, E. Okazaki, H. Kikuchi, and K. Iwanami. "Electro – pneumatically controlled hand prosthesis using pattern recognition of myo – electric signals", in *Digest of 7th ICMBE*, p. 367, 1967.
[31] R. W. Mann and S. D. Reimers. Kinesthetic sensing for EMG controlled "Boston arm." *IEEE Transactions on Man – Machine Systems*, MMS – 11:110 – 115, 1970.
[32] T. A. Rohland. Sensory feedback for powered limb prostheses. *Medical and Biological Engineering*, 12:300 – 301, March 1975.
[33] B. S. Hudgins. *A Novel Approach to Multifunctional Myoelectric Control of Prosthesis.* PhD Dissertation, University of New Brunswick, Fredericton, Canada, 1991.
[34] C. M. Light, P. H. Chappell, B. Hudgins, and K. Engelhart. Intelligent mul – tifunction myoelectric control of hand prostheses. *Journal of Medical Engineering and Technology*, 26(4):139 – 146, July – August 2002.
[35] H. Hanafusa and H. Asada. "Stable prehension by a robot hand with elastic fingers", in M. Brady, J. M. Hollerbach, T. L. Johnson, T. Lozano – Perez, and M. T. Mason, editors, *Robot Motion: Planning and Control*, pp. 322 – 335. MIT Press, Cambridge, Massachusetts, USA, 1977.
[36] E. F. R. Crossley and F. G. Umholtz. Design of a three – fingered hand. *Mechanism and Machine Theory*, 12: 85 – 93, 1977.
[37] T. Okada. Computer control of multijointed finger system for precise object – handling. *IEEE Transaction on Systems, Man, and Cybernetics*, 12(3):289 – 299, 1982.
[38] S. C. Jacobsen, D. F. Knutti, R. T. Johnson, and H. H. Sears. Development of the Utah artificial arm. *IEEE Transactions on Biomedical Engineering*, BME – 29(4):249 – 269, April 1982.
[39] S. C. Jacobsen, J. E. Wood, D. F. Knutti, K. B. Biggers, and E. K. Iversen. The version I Utah/MIT dextrous hand. In H. Hanafusa and H. Inoue, editors, *Robotics Research: The Second International Symposium*, pp. 301 – 308. MIT Press, Cambridge, Massachusetts, USA, 1985.
[40] E. Iversen, H. H. Sears, and S. C. Jacobsen. Artificial arms evolve from robots, or vice versa? *IEEE Control*

Systems Magazine,25(1):16 – 18,20,February 2005.

[41] J. K. Salisbury. *Kinematic and Force Analysis of Articulated Hands.* PhD Dissertation, Stanford University, Stanford, California, USA, 1982.

[42] S. T. Venkataraman and T. E. Djaferis. "Multivariable feedback control of the JPL/stanford hand," in *Proceedings of the IEEE International Conference on Robotics and Automation*, pp. 77 – 82, 1987.

[43] D. Lian, S. Peterson, and M. Donath. "A three – fingered articulated hand," in *Proceedings of the 13th International Symposium on Industrial Robots*, pp. 18. 91 – 18. 101, 1983.

[44] P. J. Kyberd and J. L. Pons. "A comparison of the Oxford and Manus intelligent hand prostheses," in *Proceedings of the 2003 IEEE International Conference on Robotics and Automation*, pp. 3231 – 3236, Taipei, Taiwan, September 2003.

[45] J. L. Pons, E. Rocon, R. Ceres, D. Reynaerts, B. Saro, S. Levin, and W. Van Moorleghem. The MANUS – HAND dextrous robotics upper limb prosthesis: mechanical and manipulation aspects. *Autonomous Robots*, 16:143 – 163, 2004.

[46] H. Kobayashi. Control and geometric considerations for an articulated robot hand. *Journal of Robotic Research*, 1(1):3 – 12, 1985.

[47] A. Rovetta. "Sensors controlled multifingered robot hand," in *Proceedings of the IEEE Conference on Robotics and Automation*, pp. 1060 – 1063, St. Louis, Missouri, USA, March 1983.

[48] J. J. Kim, D. R. Blythe, D. A. Penny, and A. A. Goldenberg. "Computer archi – tecture and low level control of the PUMA/RAL – hand system," in *Proceedings of the IEEE Conference on Robotics and Automation*, pp. 1590 – 1594, Raleigh, North Carolina, USA, March 1987.

[49] H. Van Brussel, B. Santoso, and D. Reynaerts. "Design and control of a mul – tifingered hand provided with tactile feedback," in *Proceedings of the NASA Conference on Space Telerobotics*, pp. 89 – 101, Pasadena, California, USA, J anuary/February 1989.

[50] G. A. Bekey, R. Ibmovic, and I. Zeljkovic. "Control architecture for the belgrade/usc hand," in S. T. Venkataraman and T. Iberall, editors, *Dextrous Robot Hands*, pp. 136 – 149. Springer – Verlag, New York, USA, 1990.

[51] R Kyberd and R H. Chappell. The Southampton Hand: an intelligent myoelectric prosthesis. *Journal of Rehabilitation Research and Development*, 31(4):326 – 334, 1994.

[52] P. J. Kyberd, O. E. Holland, P. H. Chappell, S. Smith, R. Tregidgo, P. J. Bagwell, and M. Snaith. MARCUS: a two degree of freedom hand prosthesis with hierarchical grip control. *IEEE Transactions on Rehabilitation Engineering*, 3(1):70 – 76, March 1995.

[53] R. Okuno, M. Yoshida, and K. Akazawa. Development of biomimetic prosthetic hand controlled by electromyogram, in *1996 4th International Workshop on Advanced Motion Control*, pp. 103 – 108, Mie, Japan, March 1996.

[54] C. S. Lovchik and M. A. Diftler. "The Robonaut hand: a dexterous robot hand for space," in *Proceedings of the IEEE International Conference on Robotics and Automation*, pp. 907 – 912, May 1999.

[55] H. Huang and C. Chen. "Development of a myoelectric discrimination system for a multi – degree prosthetic," in *Proceedings of the 1999 IEEE International Conference on Robotics and Automation*, pp. 2392 – 2397, Detroit, Michigan, USA, May 1999.

[56] D. Nishikawa, W. Yii, H. Ybkoi, and Y. Kakazu. On – line learning method for EMG prosthetic hand control.

Electronics and Communications in Japan (Part III: Fundamental Electronic Science), 84 (10): 35 – 46, 2001 (Translated from Denshi Joho Tsushin Gakkai Ronbunshi, Vol. J82 – D – II, No. 9, September 1999, pp. 1510 – 1519).

[57] H. Liu, J. Butterfass, S. Knoch, R Meusel, and G. Hirzinger. A new control strategy for DLR's multisensory articulated hand. *IEEE Control Systems Magazine*, 19(2): 47 – 54, April 1999.

[58] J. Butterfass, M. Grebenstein, H. Liu, and G. Hirzinger. "DLR – Hand II: next generation of a dextrous robot hand," in *Proceedings of the IEEE International Conference on Robotics and Automation*, pp. 109 – 114, 2001.

[59] N. Fukaya, S. Toyama, T. Asfour, and R. Diffmann. "Design of the TUAT/Karlsruhe humanoid hand," in *Proceedings of the 2000 IEEE/RSJ International Conference on Intelligent Robots and Systems*, pp. 1 – 6, Stanford, California, USA, July 1996.

[60] Y. Zhang, Z. Han, H. Zhang, X. Shang, T. Wang, and W. Guo. "Design and control of the BUAA four – fingered hand," in *Proceedings of the 2001 IEEE International Conference on Robotics and Automation*, pp. 2517 – 2522, Seoul, South Korea, May 2001.

[61] N. Dechev, W. L. Cleghorn, and S. Naumann. Multiple finger, passive adaptive grasp prosthetic hand. *Mechanism and Machine Theory*, 36: 1157 – 1173, 2001.

[62] R. Kolluru, K. P. Valavanis, P. Kimon, S. Smith, and N. Tsourveloudis. An overview of the University of Louisiana robotic gripper system project. *Trans – actions of the Institute of Measurement and Control*, 24 (1): 65 – 84, 2002.

[63] J. Yang, E. P. Pitarch, K. Abdel – Malek, A. Patrick, and L, Lindkvist. A multi – fingered hand prosthesis. *Mechanism and Machine Theory*, 39 (6): 555 – 581, June 2004.

[64] R. Suarez and P. Grosch. "Dexterous robotic hand MA – I software and hardware architecture," in *Proceedings of the Intelligent Manipulation and Grasping*, pp. 91 – 96, Genova, 2004.

[65] S. Roccella, M. C. Carrozza, G. Cappeiello, M. Zecca, H. Miwa, K. Itoh, and M. Matsumoto. "Design, fabrication and preliminary results of a novel anthropomorphic hand for humanoid robotics: RCH – 1," in *Proceedings of the 2004 IEEE/RSJ International Conference on Intelligent Robots and Systems*, pp. 266 – 271, Sendai, Japan, September 28 – October 2, 2004.

[66] F. Lotti, P. Tiezzi, G. Vassura, L. Biagiotti, G. Palli, and C. Melchiorri. "Devel – opment of UB hand 3: early results," in *Proceedings of the 2005 IEEE Inter – national Conference on Robotics and Automation*, pp. 4488 – 4493, Barcelona, Spain, April 2005.

[67] T. R. Farrell, R. F. Weir, C. W. Heckathome, and D. S. Childress. The effect of static friction and backlash on extended physiological proprioception control of a powered prosthesis. *Journal of Rehabilitation Research and Development*, 42 (3): 327 – 342, May – June 2005.

[68] B. Choi, S. Lee, H. R. Choi, and S. Kang. "Development of anthropomorphic robot hand with tactile sensor: SKKU Hand II," in *Proceedings of the 2006 IEEE/RSJ International Conference on Intelligent Robots and Systems*, pp. 3779 – 3784, Beijing, P. R. China, October 9 – 15, 2006.

[69] APL to Lead Team Developing Revolutionary Prosthesis. Press Release: Dated February 9, 2006.

[70] Revolutionizing Prosthetics 2009 Team Delivers First DARPA Limb Proto – type. Press Release: Dated April 26, 2007.

[71] D. J. Atkins, D. C. Y. Heard, and W. H. Donovan. Epidemiologic overview of individuals with upper limb loss

and their reported research priorities. *Journal of Prosthetic and Orthotics*, 8(1):2 – 11, 1996.

[72] D. S. Childress. Closed – loop control in prosthetic systems: Historical perspective. *Journal Annals of Biomedical Engineering*, 8(4 – 6):293 – 303, July 1980 (45 references).

[73] R. N. Scott and R A. Parker. Myoelectric prostheses: state of the art. *Journal of Medical Engineering and Technology*, 12(4):143 – 151, July/August 1988.

[74] R. A. Grupen, T. C. Henderson, and I. D. McMammon. A survey of general purpose manipulation. *International Journal of Robotics Research*, 8(1):38 – 62, 1989.

[75] H. H. Sears and J. Shaperman. Proportional myoelectric hand control: an evaluation. *American Journal of Physical Medicine and Rehabilitation*, 70(1):20 – 28, February 1991.

[76] K. B. Shimoga. Robot grasp synthesis algorithms: a survey. *The International Journal of Robotics Research*, 15:230 – 266, 1996 (Survey article with over 130 references).

[77] A. Bicchi. Hands for dexterous manipulation and robust grasping: a difficult road toward simplicity. *IEEE Transactions on Robotics and Automation*, 16(6):652 – 662, December 2000 (Summary article with 191 references).

[78] A. M. Okamura, N. Smaby, and M. R. Cutkosky. "4An overview of dexterous manipulation," in *Proceedings of the IEEE International C2000 conferenceon Robotics and Automation*, pp. 255 – 262, San Francisco, California, USA, April 2000 (52 references).

[79] P. Kyberd, P. H. Chappell, and D. Gow. Advances in the control of prosthetic arms: Guest Editorial. *Technology and Disability*, 15(2):57 – 61, 2003.

[80] A. Muzumdar, editor. *Powered Upper Limb Prostheses Control, Implementation and Clinical Application.* Springer – Verlag, New York, USA, 2004.

[81] D. P. J. Cotton, A. Cranny, P. H. Chappell, N. M. White, and S. P. Beeby. "Control strategies for a multiple degree of freedom prosthetic hand," in *Proceedings of The Institution of Measurement and Control UK ACC Control 2006 Symposium*, pp. 211 – 218, 2006.

[82] S. Arimoto. *Control Theory of Multi – fingered Hands: A Modeling and Analytical – Mechanics Approach for Dexterity and Intelligence.* Springer – Verlag, London, UK, 2008.

[83] L. Birglen, T. Laliberte, and C. Gosselin. *Underactuated Robotic Hands.* Springer Tracts in Advanced Robotics. Springer – Verlag, Berlin, Germany, 2008.

[84] T. Inoue and S. Hirai. *Mechanics and Control of Soft – fingered Manipulation.* Springer, New York, USA, 2009.

[85] D. S. Naidu, C. – H. Chen, A. Perez, and M. P. Schoen. "Control strategies for smart prosthetic hand technology: An overview," in *The 30th Annual International Conference of the IEEE Engineering Medicine and Biology Society (EMBS)*, pp. 4314 – 4317, Vancouver, Canada, August 20 – 24, 2008 (In *Top 20 Articles, in the Domain of Article 19163667, Since its Publication (2008)* according to *BioMedLib*: "*Who is Publishing in My Domain?*," Ranked as No. 8 of 20 as on August 1, 2014, Ranked as No. 9 of 20 as on May 4, 2015 and as on June 2, 2015).

[86] D. S. Naidu and C. – H. Chen. "Automatic control techniques for smart prosthetic hand technology: An overview," in U. R. Acharya, F. Molinari, T. Tamura, D. S. Naidu, and J. Suri, editors, *Distributed Diagnosis and Home Healthcare (D2H2): Volume 2*, pp. 201 – 223. American Scientific Publishers, Stevenson Ranch, California, USA, 2011.

[87] S. Micera, J. Carpaneto, and S. Raspopovic. Control of hand prostheses using peripheral information. *IEEE Reviews in Biomedical Engineering*, 3:48 – 68, 2010 (Review article with 161 references).

[88] A. Cloutier and J. Yang. Design, control, and sensory feedback of externally powered hand prostheses: A li-terature review. *Critical Reviews in Biomedical Engineering*, 41(2):161 – 181, 2013.

[89] E. F. Murphy and A. B. Wilson. "Limb prosthetics and orthotics," in M. Clynes and J. H. Milsum, editors, *Biomedical Engineering Systems*, pp. 489 – 549. McGraw – Hill, New York, USA, 1970.

[90] N. Wiener. *Cybernetics or Control and Communication in the Animal and the Machine.* MIT Press, Cambridge, Massachusetts, USA, 1948. Second Edition, 1961; Paper Back Edition 1965.

[91] C. K. Battye, A. Nightingale, and J. Whillis. The use of myo – electric currents in the operation of prostheses. *Journal of Bone and Joint Surgery*, 37 – B:506, 1955.

[92] A. H. Bottomley. Myo – electric control of powered prostheses. *The Journal of Bone and Joint Surgery*, 47 – B (3):411 – 415, August 1965.

[93] A. H. Bottomley, G. Kingshill, P. Robert, D. Styles, P. H. Jilbert, J. W. Birtill, and J. R. Truscott. Prosthetic hand with improved control system for activation by electromyogram signals. Tbchnical report, National Research Development Corporation, London, UK, December 1968. US Patent 3418662.

[94] R. W. Todd. *Adaptive Control of a Human Prosthesis.* PhD Dissertation, University of Southampton, Southampton, UK, 1969.

[95] G. N. Saridis and H. E. Stephanou. A hierarchical approach to the control of a prosthetic arm. *IEEE Transactions on Systems, Man and Cybernetics*, 7(6):407 – 420, June 1977.

[96] G. N. Saridis and T. R Gootee. EMG pattern analysis and classification for a prosthetic arm. *IEEE Transactions on Biomedical Engineering*, BME – 29(6):403 – 412, June 1982.

[97] S. Lee and G. N. Saridis. The control of a prosthetic arm by EMG pattern recognition. *IEEE Transactions on Automatic Control*, 29(4):290 – 302, April 1984.

[98] A. A. Cole, J. E. Hauser, and S. S. Sastry. Kinematic and control of multifin – gered hands with rolling contact. *IEEE Transactions on Automatic Control*, 34(4):398 – 404, April 1989.

[99] R. M. Murray, Z. Li, and S. S. Sastry. *A Mathematical Introduction to Robotic Manipulation.* CRC Press, Boca Raton, Florida, USA, 1994.

[100] C. J. Abu – Haj and N. Hogen. Functional assessment of control systems for cybernetic elbow prostheses – Part I: Description of the technique. *IEEE Trans – actions on Biomedical Engineering*, 37(11):1025 – 1036, November 1990.

[101] C. J. Abu – Haj, N. Hogen. Functional assessment of control systems for cy – bernetic elbow prostheses – Part II: Application of the technique. *IEEE Trans – actions on Biomedical Engineering*, 37(11):1037 – 1047, November 1990.

[102] G. R Starr. Experiments in assembly using a dexterous hand. *IEEE Transac – tions on Robotics and Automation*, 6(3):342 – 347, June 1990.

[103] A. E. Hines, N. E. Owens, and P. E. Crago. Assessment of input – output proper – ties and control of neuroprosthetic hand grasp. *IEEE Transactions on Biomed – ical Engineering*, 39(6):610 – 623, June 1992.

[104] B. S. Hudgins, P. Parker, and R. N. Scott. A new strategy for multifunc – tion myoelectric control. *IEEE Transactions on Biomedical Engineering*, 40(1):82 – 94, January 1993.

[105] D. Graupe, J. Magnussen, and A. A. Beex. A microprocessor system for multifunctional control of upper –

limb prostheses via myoelectric signal iden – tification. *IEEE Transactions on Automatic Control*, 23(4): 538 – 544, August 1978.

[106] D. Graupe, R. W. Liu, and G. S. Moschytz. "Applications of neural networks to medical signal processing," in *Proceedings of the 27th IEEE Conference on Decision and Control*, pp. 343 – 347, Austin, Texas, USA, December 1988.

[107] K. Englehart and B. Hudgins. A robust, real – time control scheme for multi – function myoelectric control. *IEEE Transactions on Biomedical Engineering*, 50(7): 848 – 854, July 2003.

[108] K. Englehart, B. Hudgins, and AID. C. Chan. Continuous multifunction myoelectric control using pattern recognition. *Technology and Disability*, 15(7): 95 – 103, 2003.

[109] T. Iberall, G. Sukhatme, D. Beattie, and G. A. Bekey. "Control philosophy and simulation of a robotic hand as a model for prosthetic hands," in *Proceedings of the 1993 IEEE/RSJ International Conference on Intelligent Robots and Systems*, pp. 824 – 831, Yokohama, Japan, July 1993.

[110] T. Iberall, G. Sukhatme, D. Beattie, and G. A. Bekey. "Control philosophy for a simulated prosthetic hand," in *Proceedings of the Rehabilitation Engineering and Assistive Technology Society of North America (RESNA)*, pp. 12 – 17, Las Vegas, Nevada, USA, June 1993.

[111] T. Iberall, G. Sukhatme, D. Beattie, and G. A. Bekey. "On the development of EMG control for a prosthesis using a robotic hand," in *Proc, of the 1994 IEEE International Conference on Robotics and Automation*, pp. 1753 – 1758, San Diego, California, USA, May 1994.

[112] D. J. Bak. Control system matches prosthesis to patient. *Design News*, p. 68, June 1997.

[113] N. Sarkar, X. Yun, and V. Kumar. Dynamic control of 3 – D rolling contacts in two – arm manipulation. *IEEE Transactions on Robotics and Automation*, 13(3): 364 – 376, June 1977.

[114] A. Tura, C. Lamberti, A. Davalii, and R. Sacchetti. Experimental development of a sensory control system for an upper limb myoelectric prosthesis. *Journal of Rehabilitation Research and Development*, 35(1): 14 – 26, January 1998.

[115] C. Bonivento and A. Davalli. Automatic tuning of myoelectric prosthesis. *Journal of Rehabilitation Research and Development*, 35(3): 294 – 310, July 1998.

[116] T. Schlegl and M. Buss. "Hybrid closed – loop control of robotic hand regrasp – ing," in *Proceedings of the 1998 IEEE International Conference on Robotics and Automation*, pp. 3026 – 3031, Leuven, Belgium, May 1998.

[117] S. Micera, A. M. Sabatini, P. Dario, and B. Rossi. A hybrid approach to EMG pattern analysis for classification of arm movements using statistical and fuzzy techniques. *Medical Engineering Physics*, 21 (5): 303 – 311, June 1999.

[118] T. Tsuji, O. Fukuda, H. Shigeyoshi, and M. Kaneko. "Bio – mimetic impedance control of and EMG – controlled prosthetic hand," in *Proceedings of the 2000 IEEE/RSJ International Conference on Intelligent Robots and Systems*, pp. 377 – 382, Takamatsu, Japan, 2000.

[119] N. Hogan. Impedance control: An approach to manipulation, Part I, Part II and Part III. *Transactions of ASME, Journal of Dynamic Systems, Measurement, and Control*, 107: 1 – 24, March 1985.

[120] A. Marigo and A. Bicchi. Rolling bodies with regular surface: Control – lability theory and applications. *IEEE Transactions on Automatic Control*, 45(9): 1586 – 1599, September 2000.

[121] S. Morita, K. Shibata, X. Z. Zheng, and K. Ito. "Prosthetic hand control based on torque estimation from

EMG signals," in *Proceedings of the 2000 IEEE/RSJ International Conference on Intelligent Robots and Systems*, pp. 389 – 394, Takamatsu, Japan, October 31 – November 5, 2000.

[122] S. Morita, T. Kondo, and K. Ito, "Estimation of forearm movement from EMG signal and application to prosthetic hand control," in *Proceedings of the 2001ICRA/IEEE International Conference on Robotics and Automation*, pp. 1477 – 1482, Seoul, South Korea, May 21 – 25, 2001.

[123] P. H. Peckham, K. L. Kilgore, M. W. Keith, A. M. Bryden, N. Bhadra, and F. W. Montague. An advanced neuroprosthesis for restoration of hand and up – per arm control using an implantable controller. *The Journal of Hand Surgery*, 27A(2): 265 – 276, March 2002.

[124] B. Massa, S. Roccella, M. C. Carrozza, and P. Dario. "Design and development of an underactuated prosthetic hand," in *Proceedings of the 2002 IEEE International Conference on Robotics and Automation*, pp. 3374 – 3379, Wasington, District of Columbia, May 2002.

[125] Blatchford intelligent prosthesis, http://www. blatchford. co. uk, 1994.

[126] C. Lake and J. M. Miguelez. Evolution of microprocessor based control systems in upper extremity prosthetics. *Technology and Disability*, 15(2): 63 – 71, 2003.

[127] M. C. Carrozza, F. Vecchi, F. Sebastiani, G. Cappiello, S. Roccella, M. Zecca, R. Lazzarini, and P. Dario. "Experimental analysis of an innovative prosthetic hand with proprioceptive sensors," in *Proceedings of the 2003 IEEE International Conference on Robotics and Automation*, pp. 2230 – 2235, September 14 – 19, 2003.

[128] P. Scherillo, B. Siciliano, L. Zollo, M. C. Carrozza, E. Guglielmelli, and R Dario. "Parallel force/position control of a novel biomechatronic hand prosthesis," in *Proceedings of the 2003 IEEE/ASME International Conference on Advanced Intelligent Mechatronics (AIM 2003)*, pp. 920 – 925, 2003.

[129] D. H. Plettenburg and J. L. Herder. Voluntary closing: a promising opening in hand prosthetics. *Technology and Disability*, 15(2): 85 – 94, 2003.

[130] H. M. Al – angari, R. F. ff. Weir, C. W. Heckanthorne, and D. S. Childress. A two – degree – of – freedom microprocessor based extended physiological propri – oception (EPP) controller for upper limb prostheses. *Technology and Disability*, 15(2): 113 – 127, 2003.

[131] H. van der Linde, C. J. Hofstad, A. C. H. Geurts, K. Postema, J. H. B. Geertzen, and J. van Limbeek. A systematic literature review of the effect of different prosthetic components of human functioning with a lower – limb prosthesis. *Journal of Rehabilitation Research and Development*, 41(4): 555 – 570, July – August 2004 (Review article with 91 references).

[132] S. Arimoto, P. T. A. Nguyen, H. – Y. Han, and Z. Doulgeri. Dynamics and control of a set of dual fingers with soft tips. *Robotica*, 18: 71 – 80, 2000.

[133] S. Arimoto. Intelligent control of multi – fingered hands. *Annual Reviews in Control*, 28: 75 – 85, 2004.

[134] J. Cui and Z. Sun. "Visual hand motion capture for guilding a dexterous hand," in *Proceedings of the Sixth IEEE International Conference on Automatic Face and Gesture Recognition (FGR'04)*, pp. 729 – 734, May 17 – 19, 2004.

[135] M. C. Carrozza, B. Massa, S. Micera, R. Lazzarini, and M. Zecca. The development of a novel prosthetic handongoing research and preliminary results. *IEEE/ASME Transactions on Mechatronics*, 7(2): 108 – 114, 2002.

[136] M. C. Carrozza, G. Cappiello, G. Stellin, F. Zaccone, F. Vechhi, S. Micera, and R Dario. "On the develop-

ment of a novel adaptive prosthetic hand with complaint joints: experimental platform and EMG control," in *Proceedingsof the IEEE/RSJ International Conference on Intelligent Robots and Systems*, pp. 1271 – 1276, August 2005.

[137] R. Ozawa, S. Arimoto, S. Nakamura, and J. – H. Bae. Control of an object with parallel surfaces by a pair of finger robots without object sensing. *IEEE Transactions on Robotics*, 21(5):965 – 976, October 2005.

[138] R. F. Weir. *Direct Muscle Attachment as a Control Input for a Position – Servo Prosthesis Controller.* PhD Dissertation, Northwestern University, Evanston, Illinois, USA, 1995.

[139] R. F. Weir, P. R. Troyk, G. DeMichele, and D. Kerns. "Technical details of the implantable myoelectric sensor (IMES) system for multifunction prosthesis control," in *Proceedings of the 25th IEEE Engineering in Medicine and Biology 27th Annual Conference*, pp. 7337 – 7340, Shanghai, P. R. China, September 2005.

[140] R. F. Weir and A. B. Ajiboye. "A multifunction prosthesis controller based on fuzzy – logic techniques," in *Proceedings of the 25th Annual International Conference of IEEE EMBS*, pp. 17 – 21, Cancun, Mexico, September 2003.

[141] A. B. Ajiboye and R. F. Weir. A heuristic fuzzy logic approach to EMG pattern recognition for multifunctional prosthesis control. *IEEE Transactions on Neural Systems and Rehabilitation Engineering*, 13(3):280 – 291, September 2005.

[142] G. S. Dhillon and K. W. Horch. Direct neural sensory feedback and control of a prosthetic arm. *IEEE Transactions on Neural Systems and Rehabilitation Engineering*, 13(4):468 – 472, December 2005.

[143] A. Kargov, T. Asfour, C. Pylatiuk, R. Oberle, H. Klosek, S. Schulz, K. Regenstein, G. Bretthauer, and R. Dillmann. "Development of an anthropomorphic hand for a mobile assistive robot," in *9th International Conference on Rehabilitation Robotics (ICORR)*, pp. 182 – 186, Chicago, Illinois, USA, June – July 2005.

[144] A. Kargov, C. Pylatiuk, J. Martin, S. Schulz, and L. Derlein. A comparison of the grip force distribution in natural hands and in prosthetic hands. *Disability And Rehabilitation*, 26(12):705 – 711, 2004.

[145] C. Pylatiuk, S. Mounier, A. Kargov, S. Schulz, and G. Bretthauer. "Progress in the development of a multifunctional hand prosthesis," in *Proceedings of the 26th Annual International Conference of the IEEE EMBS*, pp. 4260 – 4263, San Francisco, California, USA, September 1 – 5, 2004.

[146] J. Zhao, Z. Xie, L. Jiang, H. G. Cai, H. Liu, and G. Hirzinger. "Levenberg – Marquardt based neural network control for a five – fingered prosthetic hand," in *Proceedings of the 2005 IEEE International Conference on Robotics and Automation*, pp. 1 – 6, Barcelona, Spain, April 2005.

[147] L. E. Rodriguez – Cheu, A. Casals, A. Cuxart, and A. Parra. "Towards the definition of a functionality index for the quantitative evaluation of hand – prosthesis," in *2005 IEEE/RSJ International Conference on Intelligent Robots and Systems (IROS)*, pp. 541 – 546, August 2 – 6, 2005.

[148] X. – T. Le, W. – G Kim, B. – C. Kim, S. – H. Han, J. – G. Ann, and Y. – H. Ha. "Design of a flexible multifingered robotic hand with a 12 D. O. F. and its control applications," in *Proceedings of the SICE – ICASE International Joint Conference*, pp. 1 – 5, Bexco, Busan, South Korea, October 18 – 21, 2006.

[149] J. Zhao, Z. Xie, L. Jiang, H. G. Cai, H. Liu, and G. Hirzinger. "A five – fingered underactuated prosthetic hand control scheme," in *Proceedings of the First IEEE/RAS – EMBS 2006 International Conference on Biomedical Robotics and Biomechatronics*, pp. 995 – 1000, Pisa, Italy, February 20 – 22, 2006.

[150] D. W. Zhao, L. Jiang, H. Huang, M. H. Jin, H. G. Cai, and H. Liu. "Devel – opment of a multi – DOF anthropomorphic prosthetic hand," in *Proceedings of the 2006 IEEE International Conference on Robotics and*

Biomimetics, pp. 1 – 6, Kunming, P. R. China, December 17 – 20, 2006.

[151] L. E. Rodriguez – Cheu and A. Casals. "Sensing and control of a prosthetic hand with myoelectric feedback," in *Proceedings of the First IEEE/RAS – EMBS 2006 International Conference on Biomedical Robotics and Biomechatronics*, pp. 607 – 612, Pisa, Italy, February 20 – 22, 2006.

[152] L. Zollo, S. Roccella, R. Tucci, B. Siciliano, E. Guglielmelli, M. C. Car – rozza, and R Dario. "Biomechatronic design and control of an anthropo – morphic artificial hand for prosthetics and robotic applications," in *Proceed – ings of the First IEEE/RAS – EMBS 2006 International Conference on Biomedical Robotics and Biomechatronics*, pp. 402 – 407, Pisa, Italy, February 20 – 22, 2006.

[153] T. R. Farrell and R. F. Weir. The optimal controller delay for myoelectric prostheses. *IEEE Transactions on Neural Systems and Rehabilitation Engineering*, 15(1): 111 – 118, March 2007.

[154] Q. Meng, H. Wang, P. Li, L. Wang, and Z. He. "Dexterous underwater robot hand: HEU Hand II," in *Proceedings of the 2006 IEEE International Conference on Mechatronics and Automation*, pp. 1477 – 1482, Luoyang, P. R. China, June 25 – 28, 2006.

[155] S. Klug, O. von Stryk, and B. Mohl. "Design and control mechanisms for a 3 DOF bionic manipulator," in *Proceedings of the First IEEE/RAS – EMBS 2006 International Conference on Biomedical Robotics and Biomechatronics*, pp. 450M54, Pisa, Italy, February 20 – 22, 2006.

[156] B. Rohrer and S. Hulet. "A learning and control approach based on the human neuromotor system," in *Proceedings of the First IEEE/RAS – EMBS 2006 International Conference on Biomedical Robotics and Biomechatronics*, pp. 57 – 61, Pisa, Italy, February 20 – 22, 2006.

[157] J. Zajdlik. "The preliminary design and motion control of a five – fingered prosthetic hand," in *Proceedings of the 2006 International Conference on Intelligent Engineering Systems*, pp. 202 – 206, 2006.

[158] L. Zollo, S. Roccella, E. Guglielmelli, M. C. Carrozza, and P. Dario. Biomechatronic design and control of an anthropomorphic artificial hand for prosthetic and robotic applications. *IEEE/ASME Transactions on Mechatronics*, 12(4): 418 – 429, August 2007.

[159] R. – J. Wai and M. – C. Lee. Intelligent optimal control of single – link flexible robot arm. *IEEE Transactions on Industrial Electronics*, 51(1): 201 – 220, 2004.

[160] Y. Becerikli, Y. Oysal, and A. F. Konar. Trajectory priming with dynamic fuzzy networks in nonlinear optimal control. *IEEE Transactions on Neural Networks*, 15(2): 383 – 394, 2004.

[161] C. Cipriani, F. Zaccone, G. Stellin, L. Beccai, G. Cappiello, M. C. Carrozza, and R Dario. "Closed – loop controller for a bio – inspired multi – fingered underactuated prosthesis," in *Proceedings 2006 IEEE International Conference on Robotics and Automation (ICRA 2006)*, pp. 2111 – 2116, Orlando, Florida, USA, May 15 – 19, 2006.

[162] P. X. Rong, Z. J. He, C. D. Zong, and N. Liu. "Trajectory tracking of robot based on adaptive theory," in *2008 International Conference on Intelligent Computation Technology and Automation (ICICTA 2008)*, pp. 298 – 301, Changhsa, Hunan, P. R. China, October 20 – 22, 2008.

[163] J. Cai, X. Ruan, and X. Li. "Output feedback adaptive control of uncertainty robot using observer backstepping," in *2008 International Conference on Intelligent Computation Technology and Automation (ICICTA 2008)*, pp. 404 – 408, Changhsa, Hunan, P. R. China, October 20 – 22, 2008.

[164] D. Seo and M. R. Akella. Non – certainty equivalent adaptive control for robot manipulator systems. *Systems and Control Letters*, 58(4): 304 – 308, April 2009.

[165] S. Liuzzo and R Tbmei. Prosthetic hand finger control using fuzzy sliding modes. *International Journal of Adaptive Control and Signal Processing*, 23:97 – 109, 2009.

[166] C. – Y. Chen, T. – H. S. Li, Y. – C. Yeh, and C. – C. Chang. Design and implementation of an adaptive sliding – mode dynamic controller for wheeled mobile robots. *Mechatronics*, 19:156 – 166, 2009.

[167] M. Tbrabi and M. Jahed. "A novel approach for robust control of single – linkmanipulators with visco – elastic behavior," in *Tenth International Conference on Computer Modeling and Simulation (UKSIM 2008)*, pp. 685 – 690, Cambridge, UK, April 1 – 3, 2008.

[168] E. D. Engeberg and S. G. Meek. "Adaptive object slip prevention for prosthetic hands through proportional – derivative shear force feedback," in *Proceedings of the 2008 IEEE/RSJ International Conference on Intelligent Robots and Systems*, pp. 1940 – 1945, Nice, France, September 22 – 26, 2008.

[169] E. D. Engeberg and S. G. Meek. Improved grasp force sensitivity for prosthetic hands through force – derivative feedback. *IEEE Transactions on Biomedical Engineering*, 55(2):817 – 821, 2008.

[170] E. D. Engeberg and S. G. Meek. Backstepping and sliding mode control hybridized for a prosthetic hand. *IEEE Transactions on Neural Systems and Rehabilitation Engineering*, 17(1):70 – 79, 2009.

[171] E. D. Engeberg, S. G. Meek, and M. A. Minor. Hybrid force velocity sliding mode control of a prosthetic hand. *IEEE Transactions on Biomedical Engineering*, 55(5):1572 – 1581, 2008.

[172] K. Ziaei, L. Ni, and D. W. L. Wang. Qft – based design of force and contact transition controllers for a flexible link manipulator. *Control Engineering Practice*, 17:329 – 344, 2009.

[173] W. Jiang and W. Ge. "Modeling and robust control for mobile robot," in *2008 IEEE Conference on Robotics, Automation and Mechatronics (RAM2008)*, pp. 1108 – 1112, Chengdu, P. R. China, September 21 – 24, 2008.

[174] N. Vitiello, E. Cattin, S. Roccella, F. Giovacchini, F. Vecchi, M. C. Carrozza, and R Dario. "The neurarm: towards a platform for joint neuroscience experiments on human motion control theories," in *Proceedings of the 2007 IEEE/RSJ International Conference on Intelligent Robots and Systems*, pp. 1852 – 1857, San Diego, California, USA, October 29 – November 2, 2007.

[175] D. Vrabie, F. Lewis, and M. Abu – Khalaf. Biologically inspired scheme for continuous – time approximate dynamic programming. *Transactions of the Institute of Measurement and Control*, 30:207 – 223, 2008.

[176] C. A. Cruz – Villar, J. Alvarez – Gallegos, and M. G. Villarreal – Cervantes. Concurrent redesign of an underactuated robot manipulator. *Mechatronics*, 19:178 – 183, 2009.

[177] V. Duchaine, S. Bouchard, and C. M. Gosselin. Computationally efficient predictive robot control. *IEEE/ASME Transactions on Mechatronics*, 12(5):570 – 578, 2007.

[178] G. E. Fainekos, A. Girard, H. Kress – Gazit, and G. J. Pappas. Temporal logic motion planning for dynamic robots. *Automatica*, 45:343 – 352, 2009.

[179] Y. Z. Arslan, Y. Hacioglu, and N. Yagiz. Prosthetic hand finger control using fuzzy sliding modes. *Journal of Intelligent and Robotic Systems*, 52:121 – 138, 2008.

[180] K. Onozato and Y. Maeda. "Learning of inverse – dynamics and inverse – kinematics for two – link scara robot using neural networks," in *The Society of Instrument and Control Engineers (SICE) Annual Conference 2007*, pp. 1031 – 1034, Kagawa University, Takamatsu, Japan, September 17 – 20, 2007.

[181] V. Aggarwal, G. Singhal, J. He, M. H. Schieber, and N. V. Thakor. "Towards closed – loop decoding of dexterous hand movements using a virtual integration environment," in *30th Annual International IEEE Engi-*

neering in Medicine and Biology Society Conference (*EMBC 2008*) , pp. 1703 – 1706 , Vancouver, British Columbia, Canada, August 20 – 24, 2008.

[182] K. K. Tan, S. Huang, and T. H. Lee. Decentralized adaptive controller design of large – scale uncertain robotic systems. *Automatica*, 45: 161 – 166, 2009.

[183] R. Kato, H. Ybkoi, A. H. Arieta, W. Yub, and T. Arai. Mutual adaptation among man and machine by using f – MRI analysis. *Robotics and Autonomous Systems*, 57: 161 – 166, 2009.

[184] M. da G. Marcos, J. A. T. Machado, and T. – P. Azevedo – Perdicoulis. Trajectory planning of redundant manipulators using genetic algorithms. *Communications in Nonlinear Science and Numerical Simulation*, 14 (7): 2858 – 2869, July 2009.

[185] Y. Kamikawa and T. Maeno. "Underactuated five – finger prosthetic hand inspired by grasping force distribution of humans," in *Proceedings of the 2008 IEEE/RSJ International Conference on Intelligent Robots and Systems*, pp. 717 – 722, Nice, France, September 22 – 26, 2008.

[186] R. N. Khushaba, A. Al – Ani, and A. Al – Jumaily. "Swarm intelligence based dimensionality reduction for myoelectric control," in *Proceedings of the IEEE Conference on Intelligent Sensors, Sensor Networks and Information Processing*, pp. 577 – 582, Melbourne, Australia, December 3 – 6, 2007.

[187] J. · ¥ Dieulot and F. Colas. Robust pid control of a linear mechanical axis: A case study. *Mechatronics*, 19: 269 – 273, 2009.

[188] C. – Y. Chen, M. H. – M. Cheng, C. – F. Yang, and J. – S. Chen. "Robust adaptive control for robot manipulators with friction," in *The Third International Conference on Innovative Computing Information and Control* (*ICICIC' 08*), pp. 42226, Dalian, Liaoning, P. R. China, June 18 – 20, 2008.

[189] P. Huang, J. Yan, J. Yuan, and Y. Xu. "Robust control of space robot for capturing objects using optimal control method," in *Proceedings of the 2007 International Conference on Information Acquisition* (*ICIA' 07*), pp. 397 – 402, Jeju City, South Korea, July 8 – 11, 2007.

[190] A. A. Tbotoonchi, M. R. Gharib, and Y. Farzaneh. "A new approach to control of robot," in *2008 IEEE Conference on Robotics, Automation and Mechatronics* (*RAM 2008*), pp. 649 – 654, Chengdu, P. R. China, September 21 – 24, 2008.

[191] N. Yagiz and Y. Hacioglu. Robust control of a spatial robot using fuzzy sliding modes. *Mathematical and Computer Modelling*, 49: 114 – 127, 2009.

[192] A. A. G. Siqueira and M. H. Terra. Neural network – based Hqq control for fully actuated and underactuated cooperative manipulators. *Control Engineering Practice*, 17: 418 – 425, 2009.

[193] J. L. Chen and W. – D. Chang. Feedback linearization control of a two – link robot using a multi – crossover genetic algorithm. *Expert Systems with Applications*, 36: 4154 – 4159, 2009.

[194] M. Salehi, G. R. Vbssoughi, M. Vajedi, and M. Brooshaki. "Impedance control and gain tuning of flexible base moving manipulators using PSO method," in *Proceedings of the 2008 IEEE International Conference on Information and Automation*, pp. 458 – 463, Zhangjiajie, P. R. China, June 20 – 23, 2008.

[195] B. Subudhi and A. S. Morris. Soft computing methods applied to the control of a flexible robot manipulator. *Applied Soft Computing*, 9: 149 – 158, 2009.

[196] X. Wen, D. Sheng, and J. Huang. *A Hybrid Particle Swarm Optimization for Manipulator Inverse Kinematics Control, volume 5226 of Lecture Notes in Computer Science.* Springer – Verlag, Berlin, Germany, 2008.

[197] C. – H. Chen, K. W. Bosworth, M. P. Schoen, S. E. Bearden, D. S. Naidu, and A. Perez – Gracia. "A study

of particle swarm optimization on leukocyte adhesion molecules and control strategies for smart prosthetic hand," in *2008 IEEE Swarm Intelligence Symposium (IEEE SIS08)*, St. Louis, Missouri, USA, September 21 – 23, 2008.

[198] C. – H. Chen, D. S. Naidu, A. Perez – Gracia, and M. P. Schoen. "Fusion of hard and soft control techniques for prosthetic hand," in *Proceedings of the International Association of Science and Technology for Development (IASTED) International Conference on Intelligent Systems and Control (ISC 2008)*, pp. 120 – 125, Orlando, Florida, USA, November 16 – 18, 2008.

[199] C. – H. Chen, D. S. Naidu, A. Perez – Gracia, and M. P. Schoen. "A hybrid control strategy for five – fingered smart prosthetic hand," in *Joint 48th IEEE Conference on Decision and Control (CDC) and 28th Chinese Control Conference (CCC)*, pp. 5102 – 5107, Shanghai, P. R. China, December 16 – 18, 2009.

[200] D. S. Naidu. *Optimal Control Systems.* CRC Press, a Division of Taylor & Francis, Boca Raton, Florida, USA and London, UK, 2003 (A vastly expanded and updated version of this book, is under preparation for publication in 2017).

[201] C. – H. Chen, D. S. Naidu, A. Perez – Gracia, and M. P. Schoen. "A hybrid optimal control strategy for a smart prosthetic hand," in *Proceedings of the ASME 2009 Dynamic Systems and Control Conference (DSCC)*, Hollywood, California, USA, October 12 – 14, 2009 (No. DSCC2009 – 2507).

[202] C. – H. Chen and D. S. Naidu. "Optimal control strategy for two – fingered smart prosthetic hand," in *Proceedings of the International Association of Science and Technology for Development (IASTED) International Conference on Robotics and Applications (RA 2010)*, pp. 190 – 196, Cambridge, Massachusetts, USA, November 1 – 3, 2010.

[203] F. L. Lewis, S. Jagannathan, and A. Yesildirek. *Neural Network Control of Robotic Manipulators and Nonlinear Systems.* Taylor & Francis, London, UK, 1999.

[204] F. L. Lewis, D. M. Dawson, and C. T. Abdallah. *Robot Manipulators Control: Second Edition, Revised and Expanded.* Marcel Dekker, Inc., New York, USA, 2004.

[205] C. – H. Chen, D. S. Naidu, A. Perez – Gracia, and M. P. Schoen. "A hybrid adaptive control strategy for a smart prosthetic hand," in *The 31st Annual International Conference of the IEEE Engineering Medicine and Biology Society (EMBS)*, pp. 5056 – 5059, Minneapolis, Minnesota, USA, September 2 – 6, 2009.

[206] C. – H. Chen, D. S. Naidu, and M. P. Schoen. "An adaptive control strategy for a five – fingered prosthetic hand," in *The 14th World Scientific and Engineering Academy and Society (WSEAS) International Conference on Systems, Latest Trends on Systems (Volume II)*, pp. 405T10, Corfu Island, Greece, July 22 – 24, 2010.

[207] C. – H. Chen, M. P. Schoen, and K. W. Bosworth. "A condensed hybrid optimization algorithm using enhanced continuous tabu search and particle swarm optimization," in *Proceedings of the ASME 2009 Dynamic Systems and Control Conference (DSCC)*, Hollywood, California, USA, October 12 – 14, 2009 (No. DSCC2009 – 2526).

[208] F. Liu and H. Chen. "Motion control of intelligent underwater robot based on CMAC – PID," in *Proceedings of the 2008 IEEE International Conference on Information and Automation*, pp. 1308 – 1311, Zhangjiajie, P. R. China, June 20 – 23, 2008.

[209] L. A. Zadeh. Soft computing and fuzzy logic. *IEEE Software*, 11(6): 48 – 56, 1994.

[210] L. A. Zadeh. Fuzzy sets. *Information and Control*, 8: 338 – 353, 1965.

[211] L. A. Zadeh. Outline of a new approach to the analysis of complex systems and decision processes. *IEEE Transactions on Systems*, Man, and Cybernetics, 3(1): 28 - 44, 1973.

[212] L. A. Zadeh. *Possibility Theory and Soft Data Analysis, book chapter 3, to appear in a book titled "Mathematical Frontiers of the Social and Policy Sciences"*, Westview Press, Boulder, Colorado, USA, 1981.

[213] L. Magdalena. What is soft computing? revisiting possible answers. *International Journal of Computational Intelligence Systems*, 3(2): 148 - 159, June 2010.

[214] J. - S. R. Jang, C. - T. Sun, and E. Mizutani. *Neuro - Fuzzy and Soft Computing: A Computational Approach to Learning and Machine Intelligence.* Prentice Hall PTR, Upper Saddle River, New Jersey, USA, 1997.

[215] A. Tettamanzi and M. Tomassini. *Soft Computing: Integrating Evolutionary, Neural, and Fuzzy Systems.* Springer - Verlag, Berlin, Germany, 2001.

[216] S. J. Ovaska, H. F. VanLandingham, and A. Kamiya. Fusion of soft computing and hard computing in industrial applications: An overview. *IEEE Transactions on Systems, Man, and Cybernetics, Part C: Applications and Reviews*, 32(2): 72 - 79, May 2002.

[217] F. O. Karray and C. De Silva. *Soft Computing and Intelligent Systems Design: Theory, Tools and Applications.* Pearson Educational Limited, Harlow, England, UK, 2004.

[218] A. Konar. *Computational Intelligence: Principles, Techniques and Applications.* Springer - Verlag, Berlin, Germany, 2005.

[219] C. - H. Chen, K. W. Bosworth, and M. P. Schoen. "Investigation of particle swarm optimization dynamics," in *Proceedings of International Mechanical Engineering Congress and Exposition (IMECE) 2007*, Seattle, Washington, USA, November 11 - 15, 2007 (No. IMECE2007 - 41343).

[220] C. - H. Chen, K. W. Bosworth, and M. P. Schoen. "An adaptive particle swarm method to multiple dimensional problems," in *Proceedings of the International Association of Science and Technology for Development (IASTED) International Symposium on Computational Biology and Bioinformatics (CBB2008)*, pp. 260 - 265, Orlando, Florida, USA, November 16 - 18, 2008.

[221] C. I. Christodooulu and C. S. Pattichis. Unsupervised pattern recognition for the classification of EMG signals. *IEEE Transactions on Biomedical Engineering*, 46: 169 - 178, 1999.

[222] F. H. Y. Chan, Y. - S. Yang, F. K. Lam, Y. - T. Zhang, and P. A. Parker. Fuzzy EMG classification for prosthesis control. *IEEE Transactions on Rehabilitation Engineering*, 8(3): 305 - 311, 2000.

[223] J. J. Fernandez, K. A. Farry, and J. B. Cheatham. "Waveform recognition using genetic programming: The myoelectric signal recognition problem," in *Proceedings of the First Annual Conference of Genetic Programming*, pp. 1754 - 1759, 2000.

[224] K. Kim and J. E. Colgate. Haptic feedback enhances grip force control of sEMG - controlled prosthetic hands in targeted reinnervation amputees. *IEEE Transactions on Neural Systems and Rehabilitation Engineering*, 20(6): 798 - 805, 2012.

[225] E. N. Kamavuakoa, J. C. Rosenvanga, M. F. Bga, A. Smidstrupa, E. Erko - cevica, M. J. Niemeiera, W. Jensena, and D. Farinaa. Influence of the feature space on the estimation of hand grasping force from intramuscular EMG. *Biomedical Signal Processing and Control*, 8: 1 - 5, 2013.

[226] E. D. Engeberg. A physiological basis for control of a prosthetic hand. *Biomedical Signal Processing and Control*, 8: 6 - 15, 2013.

[227] D. S. Naidu. Intelligent Control Systems. Graduate Course Class Notes, 2007.

[228] J. T. Bingham and M. R Schoen. "Characterization of myoelectric signals using system identification techniques," in *Proceedings of the 2004 ASME International Mechanical Engineering Congress and Exposition (IMECE)*, pp. 123 – 128, Anaheim, California, USA, November 13 – 19, 2004.

[229] K. Duraisamy, O. Isebor, A. Perez, M. P. Schoen, and D. S. Naidu. "Kinematic synthesis for smart hand prosthesis," in *Proceedings of the First IEEE/RAS – EMBS 2006 International Conference on Biomedical Robotics and Biomechatronics*, pp. 1135 – 1140, Pisa, Italy, February 20 – 22, 2006.

[230] D. S. Naidu and V. K. Nandikolla. "Fusion of hard and soft control strategies for left ventricular ejection dynamics arising in biomedicine," in *Proceedings of the Automatic Control Conference (ACC)*, pp. 1575 – 1580, Portland, Oregon, USA, June 8 – 10, 2005.

[231] V. K. Nandikolla and D. S. Naidu. "Blood glucose regulation for diabetic mellitus using a hybrid intelligent technique," in *Proceedings of the 2005 ASME International Mechanical Engineering Congress and Exposition (IMECE)*, pp. 1 – 6, Orlando, Florida, USA, November 5 – 11, 2005.

[232] S. J. Ovaska and H. F. VanLandingham. Guest editorial special issue on fusion of soft computing and hard computing in industrial applications. *IEEE Transaction on Systems, Man, and Cybernetics, Part C: Applications and Reviews*, 32(2): 69 – 71, May 2002.

[233] B. Sick and S. J. Ovaska. "Fusion of soft and hard computing techniques: A multi – dimensional categorization scheme," in *2005 IEEE Mid – Summer Workshop on Soft Computing in Industrial Applications*, pp. 57 – 62, Espoo, Finland, June 28 – 30, 2005.

[234] S. J. Ovaska, A. Kamiya, and Y. Chen. Fusion of soft computing and hard computing: computational structures and characteristic features. *IEEE Transaction on Systems, Man, and Cybernetics, Part C: Applications and Reviews*, 36(3): 439 – 448, May 2006.

[235] J. C. K. Lai, M. P. Schoen, A. Perez – Gracia, D. S. Naidu, and S. W. Leung. Prosthetic devices: Challenges and implications of robotic implants and biological interfaces. *Proceedings of the Institute of Mechanical Engineers (IMechE), Part H: Journal of Engineering in Medicine*, 221(2): 173 – 183, January 2007 (Special Issue on Micro and Nano Technologies in Medicine: This article listed as 1 of 20 in *Top 20 Articles, in the Domain of Article 17385571, Since its Publication (2007)* according to *BioMedLib*: "Who is Publishing in My Domain?" as on March 17, 2015).

[236] C. – H. Chen, D. S. Naidu, and M. P. Schoen. Adaptive control for a five – fingered prosthetic hand with unknown mass and inertia. *World Scientific and Engineering Academy and Society (WSEAS) Journal on Systems*, 10(5): 148 – 161, May 2011.

[237] C. – H. Chen and D. S. Naidu. "Fusion of fuzzy logic and PD control for a five – fingered smart prosthetic hand," in *Proceedings of the 2011 IEEE International Conference on Fuzzy Systems (FUZZ – IEEE 2011)*, pp. 2108 – 2115, Taipei, Taiwan, June 27 – 30, 2011.

[238] C. – H. Chen and D. S. Naidu. Hybrid control strategies for a five – finger robotic hand. *Biomedical Signal Processing and Control*, 8(4): 382 – 390, July 2013.

[239] C. – H. Chen and D. S. Naidu. A modified optimal control strategy for a five – finger robotic hand. *International Journal of Robotics and Automation Technology*, 1(1): 3 – 10, November 2014.

[240] C. – H. Chen and D. S. Naidu. "Hybrid genetic algorithm PID control for a five – fingered smart prosthetic hand," in *Proceedings of the Sixth International Conference on Circuits, Systems and Signals (CSS' ll)*,

pp. 57 – 63, Vbuliagmeni Beach, Athens, Greece, March 7 – 9, 2012.

[241] K. W. Horch and G. S. Dhillon. *Neuroprosthesis: Theory and Practice*. World Scientific, River Edge, New Jersey, USA, 2004.

[242] X. Navarro, T. B. Krueger, N. Lago, S. Micera, T. Stieglitz, and P. Dario. A critical review of interfaces with the peripheral nervous system for the control of neuroprostheses and hybrid bionic systems. *Journal of the Peripheral Nervous System*, 10: 229 – 258, 2005 (Review article with over 300 references).

[243] K. W. Horch and G. S. Dhillon. "Towards a neuroprosthetic arm," in *Proceedings of the First IEEE/RAS – EMBS 2006 International Conference on Biomedical Robotics and Biomechanics*, pp. 1 – 4, Pisa, Italy, February 2006.

[244] K. Shenoy. Toward high – performance neural control of prosthetic devices. Technical report, Stanford University, Stanford, California, USA, May 2007.

[245] R. L. Hart, K. L. Kilgore, and R H. Peckham. A comparison between control methods for implanted FES hand – grasp systems. *IEEE Transactions on Rehabilitation Engineering*, 6(2): 208 – 218, June 1998.

[246] K. Oppenheim. Jess Sullivan powers robotic arms with his mind. CNN, March 23, 2006.

[247] CNN – News: Bionic arm provides hope for amputees, September 14, 2006.

[248] R Guinnessy. DARPA. joins industry, academia to build better prosthetic arms. *Physics Today*, 59(9): 24 – 25, September 2006.

[249] CNN – News: Brain chip heralds neurotech dawn, July 17, 2006.

[250] L. R. Hochberg, M. D. Serruya, G. M. Friehs, A. Mukand, M. Saleh, A. H. Caplan, A. Branner, D. Chen, R. D. Penn, and J. R Donoghue. Neuronal ensemble control of prosthetic devices by a human with tetraplegia. *Nature*, 442(13): 164 – 171, July 2006.

[251] N. Thakor. Catching brain waves in a net. *Spectrum, IEEE*, 51(9): 40 – 45, September 2014.

第 2 章　运动学与轨迹规划

本章讨论人体手部解剖学、正向运动学、反向运动学、微分运动学和轨迹规划问题。串联操纵器定义为一系列关节相连接。其中,关节的一端固定在基架上,而另一端是自由的,出于我们的目的,末端执行器称为手指尖。串行关节的最终运动由每个关节相对于前一关节的基本运动组成。因此,为了在空间中操纵对象,必须描述每个手指的指尖(末端执行器)位置。2.2 节介绍通过正向运动学推导指尖位置的过程。这用 2.3 节中的反向运动学,需要从已知指尖位置(笛卡尔空间)获得每根手指的关节角度(关节空间)。在现实生活中,每根手指的关节角度都被限制在一定范围内。在 2.3 节中,我们生成指尖的工作区。指尖的线速度、角速度和加速度由 2.4 节中的微分运动学得到。然后利用几何雅可比方程,由指尖的线速度和角速度以及指尖的加速度推导出各指尖的角速度和角加速度。最后,在控制机械手执行特定的运动任务之前,用 2.5 节中的多项式和贝塞尔曲线函数设计所需的路径。

2.1　人体的手部

图 2.1 显示了一只正常的手,由拇指(t)、食指(i)、中指(m)、无名指(r)、小指(l)和手掌组成。手腕位于前臂和手掌之间,由 8 块腕骨组成,分为两排,分别为近端腕骨(可移动)和远端腕骨(不可移动),如图 2.1(b)[1-3] 所示。一只手有 27 块骨头,包括 5 块远端指骨、4 块中部指骨、5 块近端指骨、5 块掌骨和 8 块腕骨。腕骨近端排(上)由内侧向外侧依次为舟骨、月骨、三角骨和皮鼻骨;腕骨远端排(下)由内侧向外侧依次为钩骨、头状骨、小多角骨和多角骨。一只手由 5 块掌骨和 5 根手指组成。掌骨形成一条曲线,所以手掌在静止位置是凹陷的。5 个字母分别表示 1 根拇指(t)和其他 4 根手指,即食指(i)、中指(m)、无名指(r)和小指(l)。拇指有 2 块骨头:近端指骨和远端指骨。其他指头由 3 块骨头组成,分别为近端指骨、中指骨和远端指骨。在对机械手软硬控制策略融合研究中,我们假设手掌是固定的,拇指有 2 个关节(近指骨和远指骨),其余手指有 3 个关节(近指骨、中指骨和远指骨)。

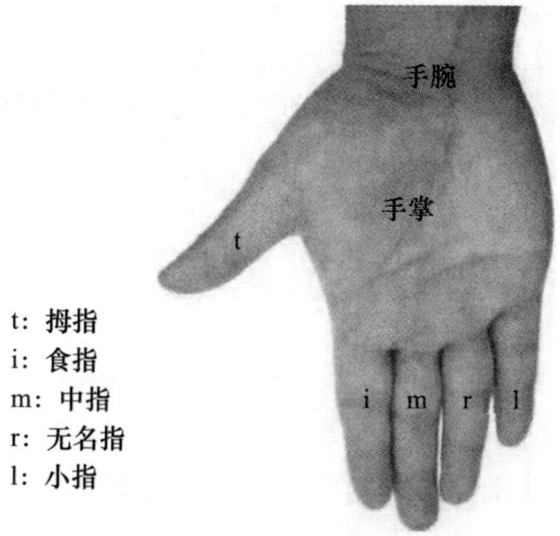

(a) 右手的外形(前视图)

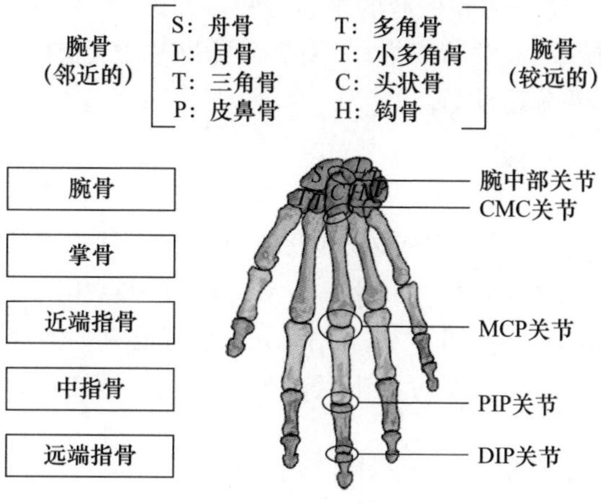

(b) 左手骨(后视图)

图 2.1　人的手腕和手

滑膜关节形成于 2 块指骨之间相对运动的表面，在拇指和其余四指的骨节处有 2 个鞍形滑膜关节面，IS4 将其定义为鞍形关节。食指、中指、无名指和小指包括 3 个旋转关节，以便进行角度运动（图 2.1（b））。掌骨指骨（MCP）关节位于掌骨和近端指骨之间；近端和远端指骨之间（PIP 和 DIP）关节分隔指骨。拇指包含 MCP 和指间（IP）关节。对于人的手，每根手指有 4 个自由度（2 个在 MCP 关节，1 个在 PIP 关节，1 个在 DIP 关节），拇指有 3 个自由度（2 个在 MCP 关节，1 个在 IP 关节），手腕有 2 个自由度，腕掌（CMC）关节有 2 个自由度。在对机械手软硬控制策略融合的研究中，模拟了 14 自由度，五指机械手有 1 根双关节拇指和4根三关节手指。q_1^j、q_2^j 和 q_3^j(j = i,m,r 和 l)分别表示食指、中指、无名指和小指的第一个关节的角位置(或关节角度)、第二个关节的角位置(或关节角度)、第三个关节的角位置(或关节角度);q_1^t 和 q_2^t 分别表示拇指(t)第一个关节的角位置 MCP^t 和第二个关节的角位置 IP^t。

2.2 前向运动

运动学是研究运动中的几何学。它仅限于运动的自然几何描述,包括位置、方向及其导数(速度和加速度)。换句话说,关节系统的正向和反向运动学研究关节的角度与末端执行器(指尖)的位置和方向之间的关系。然后,微分运动学通过操纵器的几何雅可比式表达关节角速度和角加速度与末端执行器(指尖)的线速度、角速度、线加速度、角加速度之间的解析关系。操纵器的运动学描述被用来推导机械手动力学和控制的基本方程。2.2节~2.4节将分别通过多关节串行连接的机械手、双关节拇指、三关节手指和三维五指机械手介绍正向运动学、反向运动学和差分运动学。

2.2.1 齐次变换

在引出正向运动学之前,有必要研究旋转矩阵、平移矢量和齐次变换[4]。图2.2显示了两个坐标系 $x-y-z$ 和 $x'-y'-z'$,其原点相同,围绕 z 轴以 γ 角相互旋转。设 $\boldsymbol{p}=[p_x,p_y,p_z]$ 和 $\boldsymbol{p}'=[p'_x,p'_y,p'_z]$ 分别是两个坐标系中点 P 的坐标矢量。根据几何关系,两个坐标系中点 P 坐标对应矢量之间的关系表示为

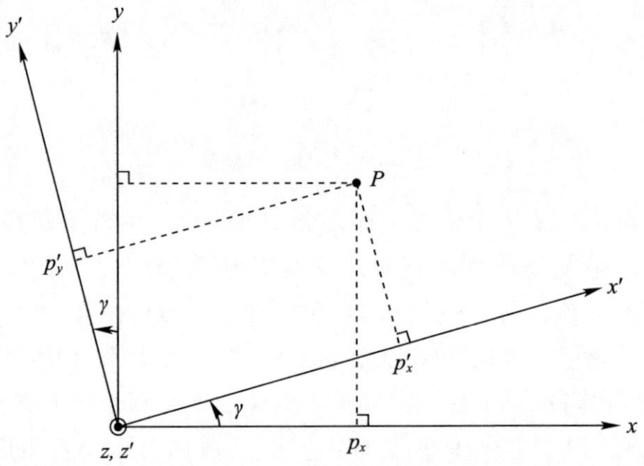

图2.2 在绕 z 轴转动 γ 角的旋转坐标系中点 P 的表示

$$
\begin{cases}
p_x = p_x' \cos\gamma - p_y' \sin\gamma \\
p_y = p_x' \sin\gamma + p_y' \cos\gamma \\
p_z = p_z'
\end{cases}
\tag{2.2.1}
$$

式(2.2.1)以矩阵形式表示如下：

$$
\begin{cases}
\begin{bmatrix} p_x \\ p_y \\ p_z \end{bmatrix} = \begin{bmatrix} \cos\gamma & -\sin\gamma & 0 \\ \sin\gamma & \cos\gamma & 0 \\ 0 & 0 & 1 \end{bmatrix} \begin{bmatrix} p_x' \\ p_y' \\ p_z' \end{bmatrix} \\
\boldsymbol{p} = \boldsymbol{R}_z(\gamma)\boldsymbol{p}'
\end{cases}
\tag{2.2.2}
$$

式中：$\boldsymbol{R}_z(\gamma)$为坐标系 $x'-y'-z'$相对于坐标系 $x-y-z$ 的旋转矩阵，其绕 z 轴旋转的角度为 γ，并有

$$
\boldsymbol{R}_z(\gamma) = \begin{bmatrix} \cos\gamma & -\sin\gamma & 0 \\ \sin\gamma & \cos\gamma & 0 \\ 0 & 0 & 1 \end{bmatrix}
\tag{2.2.3}
$$

用类似方法，可以得到坐标系 $x'-y'-z'$相对于坐标系 $x-y-z$ 的旋转矩阵 $\boldsymbol{R}_x(\alpha)$和 $\boldsymbol{R}_y(\beta)$，其中绕 x 轴和 y 轴旋转的角度为 α 和 β，且

$$
\boldsymbol{R}_x(\alpha) = \begin{bmatrix} 1 & 0 & 0 \\ 0 & \cos\alpha & \sin\alpha \\ 0 & \sin\alpha & \cos\alpha \end{bmatrix}
\tag{2.2.4}
$$

$$
\boldsymbol{R}_y(\beta) = \begin{bmatrix} \cos\beta & 0 & \sin\beta \\ 0 & 1 & 0 \\ -\sin\beta & 0 & \cos\beta \end{bmatrix}
\tag{2.2.5}
$$

图 2.3 表示了空间中的任意点 P。在两个不同的坐标系 $O^0-x^0y^0z^0$ 和 $O^1-x^1y^1z^1$ 中，点 P 的位置矢量分别记为 $\boldsymbol{p}^0$ 和 $\boldsymbol{p}^1$。假设${}^0\boldsymbol{o}_1$ 和${}^0\boldsymbol{R}_1{}^0$分别是坐标系 1 相对于坐标系 0 的原点平移矢量和旋转矩阵，则点 P 相对于坐标系 $O^0-x^0y^0z^0$ 的位置矢量 $\boldsymbol{p}^0$ 可以写为

$$
\boldsymbol{p}_0 = {}^0\boldsymbol{o}_1 + {}^0\boldsymbol{R}_1\boldsymbol{p}^1
\tag{2.2.6}
$$

式(2.2.6)是两坐标系中平移矢量${}^0\boldsymbol{o}_1$ 和旋转矩阵${}^0\boldsymbol{R}_1$ 的坐标变换。坐标系

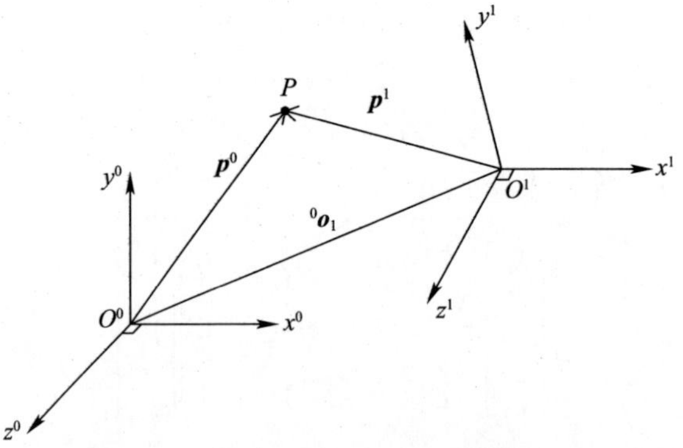

图 2.3　点 P 在两个不同坐标系中的表示

1 对坐标系 0 的齐次变换矩阵 ${}^0\boldsymbol{T}_1$ 用 4×4 矩阵表示为

$$
{}^0\boldsymbol{T}_1 = \left[\frac{{}^0\boldsymbol{R}_1(3\times3)\,|\,{}^0\boldsymbol{o}_1(3\times1)}{0(1\times3)\,|\,1(1\times3)}\right]
$$

$$
= \left[\frac{\text{旋转}\,|\,\text{变换}}{\text{角度}\,|\,\text{比例因子}}\right] \tag{2.2.7}
$$

因此,在坐标系 0 和 1 中,点 P 的齐次位置矢量用齐次变换矩阵表示为

$$
\tilde{\boldsymbol{p}}^0 = {}^0\boldsymbol{T}_1\tilde{\boldsymbol{p}}^1 \tag{2.2.8}
$$

式中:$\tilde{\boldsymbol{p}}^0 = [\boldsymbol{p}^0 \quad 1]$;$\tilde{\boldsymbol{p}}^1 = [\boldsymbol{p}^1 \quad 1]'$,(′)表示矢量或矩阵的转置。

图 2.4 显示了旋转矩阵的齐次变换 $\boldsymbol{R}(x,\alpha)$、$\boldsymbol{R}(y,\beta)$ 和 $\boldsymbol{R}(z,\gamma)$,旋转矩阵的旋转角分别为 α、β 和 γ(围绕 x 轴、y 轴和 z 轴旋转)。

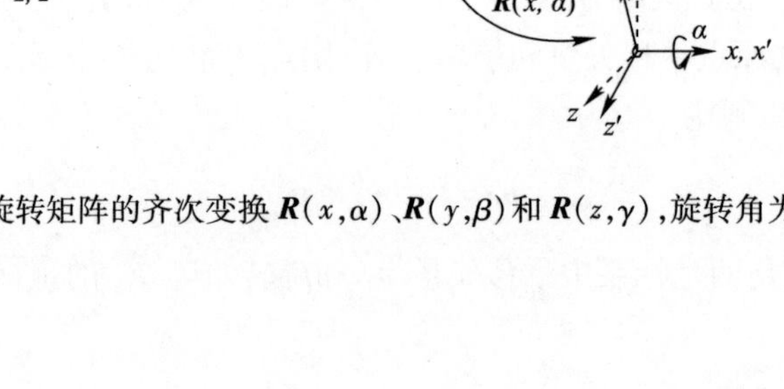

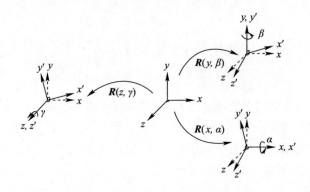

图 2.4　旋转矩阵的齐次变换 $\boldsymbol{R}(x,\alpha)$、$\boldsymbol{R}(y,\beta)$ 和 $\boldsymbol{R}(z,\gamma)$,旋转角为 α、β 和 γ

$\boldsymbol{R}(x,\alpha)$、$\boldsymbol{R}(y,\beta)$和$\boldsymbol{R}(z,\gamma)$可写为

$$\begin{cases} \boldsymbol{R}(x,\alpha)=\begin{bmatrix} 1 & 0 & 0 & 0 \\ 0 & \cos\alpha & -\sin\alpha & 0 \\ 0 & \sin\alpha & \cos\alpha & 0 \\ 0 & 0 & 0 & 1 \end{bmatrix} \\ \boldsymbol{R}(y,\beta)=\begin{bmatrix} \cos\beta & 0 & \sin\beta & 0 \\ 0 & 1 & 0 & 0 \\ -\sin\beta & 0 & \cos\beta & 0 \\ 0 & 0 & 0 & 1 \end{bmatrix} \\ \boldsymbol{R}(z,\gamma)=\begin{bmatrix} \cos\gamma & -\sin\gamma & 0 & 0 \\ \sin\gamma & \cos\gamma & 0 & 0 \\ 0 & 0 & 1 & 0 \\ 0 & 0 & 0 & 1 \end{bmatrix} \end{cases} \tag{2.2.9}$$

与之类似,图 2.5 显示了平移矢量的齐次变换 $\boldsymbol{D}(x,a)$、$\boldsymbol{D}(y,b)$和 $\boldsymbol{D}(z,c)$分别相对于 x 轴、y 轴和 z 轴位移 a、b 和 c 的情况。$\boldsymbol{D}(x,a)$、$\boldsymbol{D}(y,b)$和 $\boldsymbol{D}(z,c)$写为

$$\begin{cases} \boldsymbol{D}(x,a)=\begin{bmatrix} 1 & 0 & 0 & a \\ 0 & 1 & 0 & 0 \\ 0 & 0 & 1 & 0 \\ 0 & 0 & 0 & 1 \end{bmatrix} \\ \boldsymbol{D}(y,b)=\begin{bmatrix} 1 & 0 & 0 & 0 \\ 0 & 1 & 0 & b \\ 0 & 0 & 1 & 0 \\ 0 & 0 & 0 & 1 \end{bmatrix} \\ \boldsymbol{D}(z,c)=\begin{bmatrix} 1 & 0 & 0 & 0 \\ 0 & 1 & 0 & 0 \\ 0 & 0 & 1 & c \\ 0 & 0 & 0 & 1 \end{bmatrix} \end{cases} \tag{2.2.10}$$

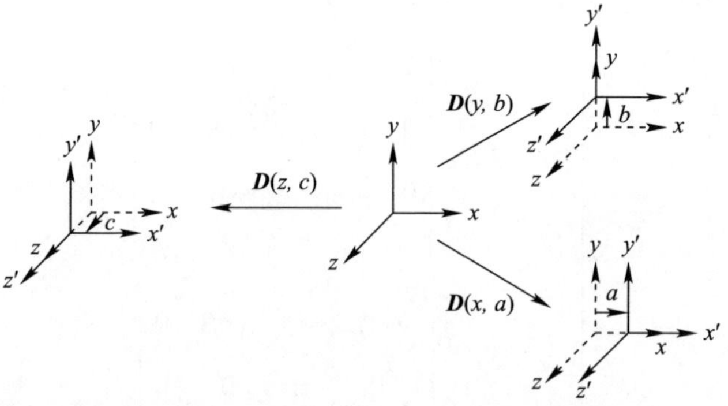

图 2.5　平移矢量的齐次变换 $\boldsymbol{D}(x,a)$、$\boldsymbol{D}(y,b)$ 和 $\boldsymbol{D}(z,c)$，位移为 a、b 和 c

2.2.2　多旋转关节串行连接的二维机械手

如图 2.6 所示，坐标系 n 相对于坐标系 0 的齐次变换矩阵 ${}^{0}\boldsymbol{T}_{n}$ 被视为 n 次坐标系变换的组合，${}^{0}\boldsymbol{T}_{n}$ 可表示为

$$ {}^{0}\boldsymbol{T}_{n} = {}^{0}\boldsymbol{T}_{1}\,{}^{1}\boldsymbol{T}_{2}\cdots{}^{n-1}\boldsymbol{T}_{n} \tag{2.2.11} $$

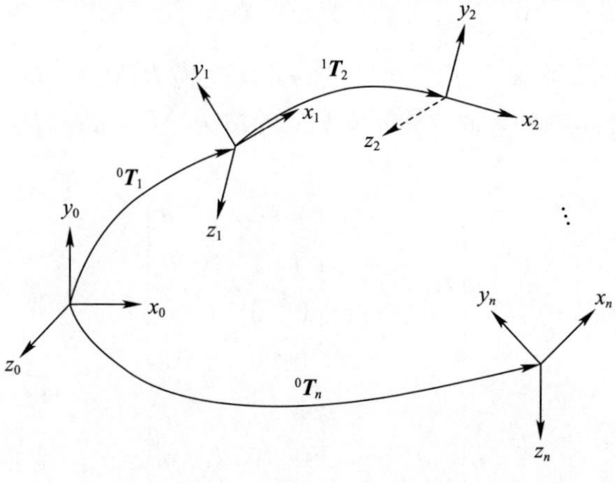

图 2.6　坐标系 n 相对于坐标系 0 的齐次变换矩阵 ${}^{0}\boldsymbol{T}_{n}$

基于式(2.2.11)中的递归表达式，需要推导出一种通用的方法来定义两个连续关节的相对位置和方向，以便计算出一个多旋转关节串行连接机械手的正向运动学方程[5]。因此，需要计算连接第$(i-1)$个和第 i 个关节的坐标系之间的变换矩阵 ${}^{i-1}\boldsymbol{T}_{i}$。

图 2.7 是一个多旋转关节串行连接的平面机械手示意图。图中，L_i 是关节长度，q_i 是关节角度$(i=1,2,\cdots,n)$。Denavit - Hartenberg (DH)方法[6-10]用于

定义第 i 个坐标系相对于第 $(i-1)$ 个坐标系的位置。DH 坐标系由 4 个参数：a_i、α_i、d_i 和 θ_i。其中：a_i 是关节 i 的运动长度，α_i 是关节 i 的扭转角；d_i 是关节偏移（或称为关节距离），它是两个关节轴之间的距离；θ_i 是关节角度。一种方便的零点配置方法是考虑所有关节沿 x 轴的排列情况。因此，根据 DH 定理，从第 $(i-1)$ 个坐标系到第 i 个坐标系的坐标变换矩阵 ${}^{i-1}\boldsymbol{T}_i$ 可以表示为 4 个基本齐次变换 $\boldsymbol{D}(z_{i-1},d_i)$、$\boldsymbol{R}(z_{i-1},\theta_i)$、$\boldsymbol{D}(x_{i-1},a_i)$ 和 $\boldsymbol{R}(x_{i-1},a_i)$ 的乘积。

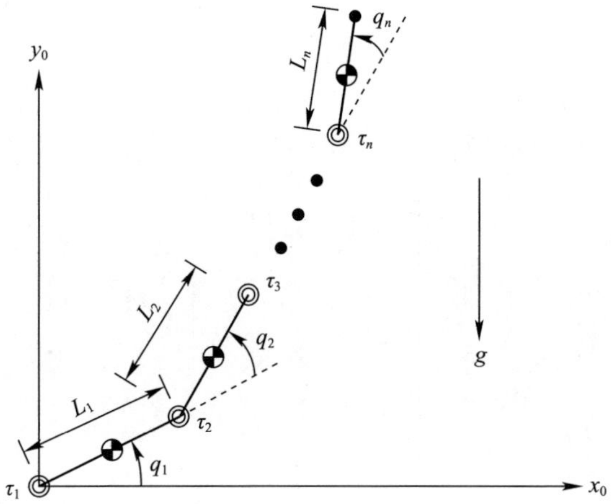

图 2.7　多旋转关节串行连接的平面机械手示意图

$$
\begin{aligned}
{}^{i-1}\boldsymbol{T}_i &= \boldsymbol{D}(z_{i-1},d_i)\boldsymbol{R}(z_{i-1},\theta_i)\boldsymbol{D}(x_{i-1},a_i)\boldsymbol{R}(x_{i-1},a_i) \\
&= \begin{bmatrix} \cos\theta_i & \sin\theta_i & 0 & -a_i \\ -\sin\theta_i\cos\alpha_i & \cos\theta_i\cos\alpha_i & \sin\alpha_i & -d_i\sin\alpha_i \\ \sin\theta_i\sin\alpha_i & -\cos\theta_i\sin\alpha_i & \cos\alpha_i & -d_i\cos\alpha_i \\ 0 & 0 & 0 & 1 \end{bmatrix}
\end{aligned}
\tag{2.2.12}
$$

$$
\boldsymbol{D}(z_{i-1},d_i) = \begin{bmatrix} 1 & 0 & 0 & 0 \\ 0 & 1 & 0 & 0 \\ 0 & 0 & 1 & d_i \\ 0 & 0 & 0 & 1 \end{bmatrix}, \boldsymbol{R}(z_{i-1},\theta_i) = \begin{bmatrix} \cos\theta_i & -\sin\theta_i & 0 & 0 \\ \sin\theta_i & \cos\theta_i & 0 & 0 \\ 0 & 0 & 1 & 0 \\ 0 & 0 & 0 & 1 \end{bmatrix}
$$

$$
\boldsymbol{D}(x_{i-1},a_i) = \begin{bmatrix} 1 & 0 & 0 & a_i \\ 0 & 1 & 0 & 0 \\ 0 & 0 & 1 & 0 \\ 0 & 0 & 0 & 1 \end{bmatrix}, \boldsymbol{R}(x_{i-1},\alpha_i) = \begin{bmatrix} 1 & 0 & 0 & 0 \\ 0 & \cos\alpha_i & -\sin\alpha_i & 0 \\ 0 & \sin\alpha_i & \cos\alpha_i & 0 \\ 0 & 0 & 0 & 1 \end{bmatrix}
\tag{2.2.13}
$$

类似于式(2.2.8),由坐标系 $i=[x_i,y_i,z_i]'$,转换到前坐标系 $(i-1)=[x_{i-1},y_{i-1},z_{i-1}]'$ 的变换方程是

$$\begin{bmatrix} x_{i-1} \\ y_{i-1} \\ z_{i-1} \\ 1 \end{bmatrix} = {}^{i-1}\boldsymbol{T}_i \begin{bmatrix} x_i \\ y_i \\ z_i \\ 1 \end{bmatrix} \tag{2.2.14}$$

从前一坐标系 $(i-1)=[x_{i-1},y_{i-1},z_{i-1}]'$ 到坐标系 $i=[x_i,y_i,z_i]'$ 的变换矩阵 ${}^{i}\boldsymbol{T}_{i-1}$ 是由 ${}^{i-1}\boldsymbol{T}_i$ 取倒数得到的。

$$\begin{bmatrix} x_i \\ y_i \\ z_i \\ 1 \end{bmatrix} = {}^{i}\boldsymbol{T}_{i-1} \begin{bmatrix} x_{i-1} \\ y_{i-1} \\ z_{i-1} \\ 1 \end{bmatrix} \tag{2.2.15}$$

$$\begin{aligned} {}^{i}\boldsymbol{T}_{i-1} &= {}^{i-1}\boldsymbol{T}_i^{-1} \\ &= \begin{bmatrix} \cos\theta_i & \sin\theta_i & 0 & -a_i \\ -\sin\theta_i\cos\alpha_i & \cos\theta_i\cos\alpha_i & \sin\alpha_i & -d_i\sin\alpha_i \\ \sin\theta_i\sin\alpha_i & -\cos\theta_i\sin\alpha_i & \cos\alpha_i & -d_i\cos\alpha_i \\ 0 & 0 & 0 & 1 \end{bmatrix} \end{aligned} \tag{2.2.16}$$

表 2.1 是图 2.7 所示的多旋转关节串行连接平面机械手的 DH 参数表。

表 2.1 多关节平面机械手的 DH 参数表(见图 2.7)

关节序号	a_i	α_i	d_i	θ_i
0	0	0	0	0
0	L_1	0	0	q_1
2	L_2	0	0	q_2
⋮	⋮	⋮	⋮	⋮
i	L_i	0	0	q_i
⋮	⋮	⋮	⋮	⋮
n	L_n	0	0	q_n

使用式(2.2.12)中的参数,坐标变换矩阵 ${}^{i-1}\boldsymbol{T}_i$ 将坐标系 i 转换为坐标系 $(i-1)$ 的计算公式如下:

$$
{}^{i-1}\boldsymbol{T}_i = \begin{bmatrix} \cos q_i & -\sin q_i & 0 & L_i\cos q_i \\ \sin q_i & \cos q_i & 0 & L_i\sin q_i \\ 0 & 0 & 1 & 0 \\ 0 & 0 & 0 & 1 \end{bmatrix} \tag{2.2.17}
$$

将式(2.2.17)代入式(2.2.11)和式(2.2.12)得

$$
\begin{aligned}
{}^0\boldsymbol{T}_n &= {}^0\boldsymbol{T}_1\,{}^1\boldsymbol{T}_2\cdots{}^{i-1}\boldsymbol{T}_i\cdots{}^{n-1}\boldsymbol{T}_n \\
&= \begin{bmatrix} C_{12\cdots n} & -S_{12\cdots n} & 0 & \sum\limits_{i=1}^{n} L_i C_{12\cdots i} \\ S_{12\cdots n} & C_{12\cdots n} & 0 & \sum\limits_{i=1}^{n} L_i S_{12\cdots i} \\ 0 & 0 & 1 & 0 \\ 0 & 0 & 0 & 1 \end{bmatrix}
\end{aligned} \tag{2.2.18}
$$

$$
\begin{cases}
{}^0\boldsymbol{T}_1 = \begin{bmatrix} \cos q_1 & -\sin q_1 & 0 & L_1\cos q_1 \\ \sin q_1 & \cos q_1 & 0 & L_1\sin q_1 \\ 0 & 0 & 1 & 0 \\ 0 & 0 & 0 & 1 \end{bmatrix} \\
{}^0\boldsymbol{T}_2 = \begin{bmatrix} \cos q_2 & -\sin q_2 & 0 & L_2\cos q_2 \\ \sin q_2 & \cos q_2 & 0 & L_2\sin q_2 \\ 0 & 0 & 1 & 0 \\ 0 & 0 & 0 & 1 \end{bmatrix} \\
{}^{n-1}\boldsymbol{T}_n = \begin{bmatrix} \cos q_n & -\sin q_n & 0 & L_n\cos q_n \\ \sin q_n & \cos q_n & 0 & L_n\sin q_n \\ 0 & 0 & 1 & 0 \\ 0 & 0 & 0 & 1 \end{bmatrix}
\end{cases} \tag{2.2.19}
$$

$$
\begin{cases} S_{i\cdots j} = \sin(q_i + \cdots + q_j) \\ C_{i\cdots j} = \cos(q_i + \cdots + q_j) \end{cases} \tag{2.2.20}
$$

因此,末端执行器的位置 $P_n(X_n, Y_n)$ 和方向 $\boldsymbol{\Phi}_n$ 为

$$
\begin{cases}
X_n = \sum\limits_{i=1}^{n} L_i C_{12\cdots i} = \sum\limits_{j=1}^{n} L_j \cos\left(\sum\limits_{i=1}^{j} q_i\right) \\
Y_n = \sum\limits_{i=1}^{n} L_i S_{12\cdots i} = \sum\limits_{j=1}^{n} L_j \sin\left(\sum\limits_{i=1}^{j} q_i\right) \\
\boldsymbol{\Phi}_n = \sum\limits_{i=1}^{n} q_i
\end{cases} \tag{2.2.21}
$$

2.2.3 双关节拇指

关节的运动学建模为刚体，而刚体位移特性在运动学中占据核心地位[9]。如图 2.8 所示，假定拇指（t）为双关节手指，其他四根手指包括食指（i）、中指（m）、无名指（r）和小指（l），为三关节手指（2.1 节）。

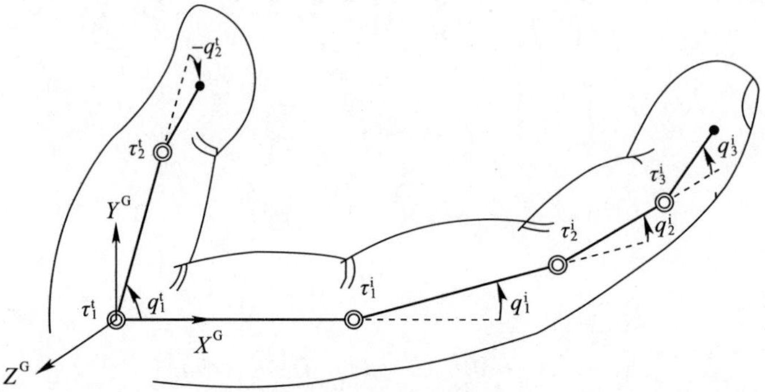

图 2.8　拇指和食指示意图

图 2.9 为双关节拇指。L_1^t 和 L_2^t 分别是拇指（t）的关节 1 和关节 2 的长度；q_1^t 和 q_1^t 分别是拇指的关节 1 和关节 2 的角度[11]。使用 DH 方法[6-7,9-10]通过 DH 变换矩阵得到拇指指尖（末端执行器）的坐标 $P^t(X^t, Y^t)$。

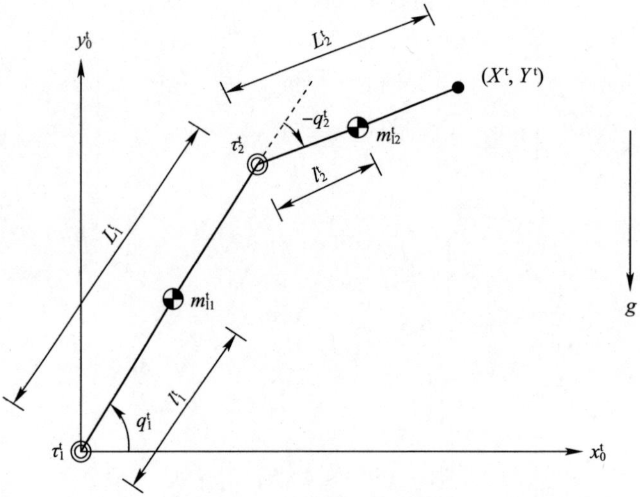

图 2.9　双关节拇指示意图

表 2.2 是图 2.9 所示双关节拇指的 DH 参数表。

表 2.2　图 2.9 所示双关节拇指的 DH 参数表

关节序号	a_i	α_i	d_i	θ_i
0	0	0	0	0
1	L_1^t	0	0	q_1^t
2	L_2^t	0	0	q_2^t

根据表 2.2,变换矩阵${}^{G}\boldsymbol{T}_0^t$、${}^{0}\boldsymbol{T}_1^t$和${}^{1}\boldsymbol{T}_2^t$可写为

$$\begin{cases} {}^{G}\boldsymbol{T}_0^t = \begin{bmatrix} 1 & 0 & 0 & 0 \\ 0 & 1 & 0 & 0 \\ 0 & 0 & 1 & 0 \\ 0 & 0 & 0 & 1 \end{bmatrix} \\ {}^{0}\boldsymbol{T}_1^t = \begin{bmatrix} \cos(q_1^t) & -\sin(q_1^t) & 0 & L_1^t\cos(q_1^t) \\ \sin(q_1^t) & \cos(q_1^t) & 0 & L_1^t\sin(q_1^t) \\ 0 & 0 & 1 & 0 \\ 0 & 0 & 0 & 1 \end{bmatrix} \\ {}^{1}\boldsymbol{T}_2^t = \begin{bmatrix} \cos(q_2^t) & -\sin(q_2^t) & 0 & L_2^t\cos(q_2^t) \\ \sin(q_2^t) & \cos(q_2^t) & 0 & L_2^t\sin(q_2^t) \\ 0 & 0 & 1 & 0 \\ 0 & 0 & 0 & 1 \end{bmatrix} \end{cases} \tag{2.2.22}$$

式中:${}^{G}\boldsymbol{T}_0^t$是从拇指的局部基准坐标系(0)到全局坐标系(G)的变换矩阵;${}^{0}\boldsymbol{T}_1^t$和${}^{1}\boldsymbol{T}_2^t$分别表示从坐标系 1 到基准坐标系(0)和从坐标系 2 到坐标系 1 的变换矩阵。因此,从拇指的局部坐标系 2 到全局坐标系的变换矩阵${}^{G}\boldsymbol{T}_2^t$写为

$$\begin{aligned} {}^{G}\boldsymbol{T}_2^t &= {}^{G}\boldsymbol{T}_0^t\,{}^{0}\boldsymbol{T}_1^t\,{}^{1}\boldsymbol{T}_2^t \\ &= \begin{bmatrix} \cos(q_1^t+q_2^t) & -\sin(q_1^t+q_2^t) & 0 & L_1^t\cos(q_1^t)+L_2^t\cos(q_1^t+q_2^t) \\ \sin(q_1^t+q_2^t) & \cos(q_1^t+q_2^t) & 0 & L_1^t\sin(q_1^t)+L_2^t\sin(q_1^t+q_2^t) \\ 0 & 0 & 1 & 0 \\ 0 & 0 & 0 & 1 \end{bmatrix} \end{aligned} \tag{2.2.23}$$

因此,拇指的指尖坐标 $P^t(X^t,Y^t)$ 和指尖方向 $\boldsymbol{\Phi}^t$ 可描述为

$$\begin{cases} X^t = L_1^t\cos(q_1^t) + L_1^t\cos(q_1^t + q_2^t) \\ Y^t = L_1^t\sin(q_1^t) + L_1^t\sin(q_1^t + q_2^t) \\ \boldsymbol{\Phi}^t = q_1^t + q_2^t \end{cases} \tag{2.2.24}$$

2.2.4 三关节食指

图 2.10 是三关节食指的示意图。图中:d 是全局坐标系(G)与食指(i)局部基准坐标系(0)之间的距离;L_1^i、L_2^i 和 L_3^i 分别是食指(i)的关节 1、2 和 3 的长度;q_1^i、q_2^i 和 q_3^i 分别是食指关节 1、2 和 3 的角度[12]。同样地,使用 DH 方法[6-7,9-10]通过 DH 变换矩阵得到食指指尖(末端执行器)的坐标 $P^i(X^i,Y^i)$。

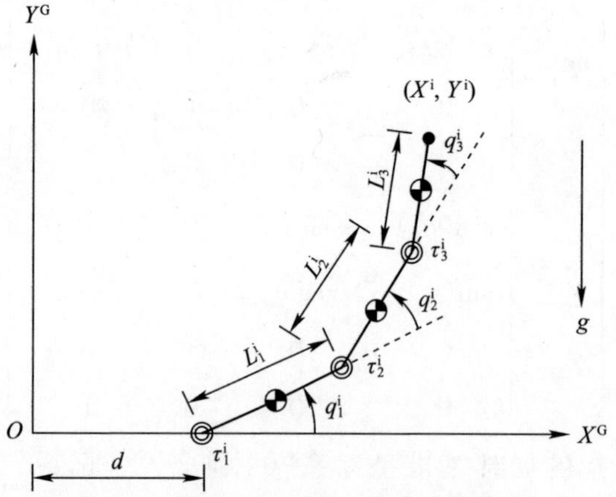

图 2.10　三关节食指示意图

表 2.3 是图 2.10 所示三关节食指的 DH 参数表。

表 2.3　三关节食指的 DH 参数表(如图 2.10)

关节序号	a_i	α_i	d_i	θ_i
0	d	0	0	0
1	L_1^i	0	0	q_1^i
2	L_2^i	0	0	q_2^i
3	L_3^i	0	0	q_3^i

根据表 2.3,变换矩阵 ${}^G\boldsymbol{T}_0^i$、${}^0\boldsymbol{T}_1^i$、${}^1\boldsymbol{T}_2^i$ 和 ${}^2\boldsymbol{T}_3^i$ 可写为

$$
\begin{cases}
{}^{G}\boldsymbol{T}_0^i = \begin{bmatrix} 1 & 0 & 0 & d \\ 0 & 1 & 0 & 0 \\ 0 & 0 & 1 & 0 \\ 0 & 0 & 0 & 1 \end{bmatrix} \\
{}^{0}\boldsymbol{T}_1^i = \begin{bmatrix} \cos(q_1^i) & -\sin(q_1^i) & 0 & L_1^i\cos(q_1^i) \\ \sin(q_1^i) & \cos(q_1^i) & 0 & L_1^i\sin(q_1^i) \\ 0 & 0 & 1 & 0 \\ 0 & 0 & 0 & 1 \end{bmatrix} \\
{}^{1}\boldsymbol{T}_2^i = \begin{bmatrix} \cos(q_2^i) & -\sin(q_2^i) & 0 & L_2^i\cos(q_2^i) \\ \sin(q_2^i) & \cos(q_2^i) & 0 & L_2^i\sin(q_2^i) \\ 0 & 0 & 1 & 0 \\ 0 & 0 & 0 & 1 \end{bmatrix} \\
{}^{2}\boldsymbol{T}_3^i = \begin{bmatrix} \cos(q_3^i) & -\sin(q_3^i) & 0 & L_3^i\cos(q_3^i) \\ \sin(q_3^i) & \cos(q_3^i) & 0 & L_3^i\sin(q_3^i) \\ 0 & 0 & 1 & 0 \\ 0 & 0 & 0 & 1 \end{bmatrix}
\end{cases}
\tag{2.2.25}
$$

式中：${}^{G}\boldsymbol{T}_0^i$是从食指(i)局部基准坐标系(0)到全局坐标系(G)的变换矩阵；${}^{0}\boldsymbol{T}_1^i$、${}^{1}\boldsymbol{T}_2^i$和${}^{2}\boldsymbol{T}_3^i$分别是从坐标系 1 到基准坐标系、从坐标系 2 到坐标系 1 以及从坐标系 3 到坐标系 2 的变换矩阵。因此，从食指局部坐标系 3 到全局坐标系的变换矩阵${}^{G}\boldsymbol{T}_3^i$可以写为

$$
\begin{aligned}
{}^{G}\boldsymbol{T}_3^i &= {}^{G}\boldsymbol{T}_0^{i0}\boldsymbol{T}_1^{i1}\boldsymbol{T}_2^{i2}\boldsymbol{T}_3^i \\
&= \begin{bmatrix} C_{123}^i & -S_{123}^i & 0 & d + L_1^iC_1^i + L_2^iC_{12}^i + L_3^iC_{123}^i \\ S_{123}^i & C_{123}^i & 0 & L_1^iS_1^i + L_2^iS_{12}^i + L_3^iS_{123}^i \\ 0 & 0 & 1 & 0 \\ 0 & 0 & 0 & 1 \end{bmatrix}
\end{aligned}
\tag{2.2.26}
$$

为了简单，我们使用了符号 $C_1^i = \cos(q_1^i)$，$S_1^i = \sin(q_1^i)$，$C_{12}^i = \cos(q_1^i + q_2^i)$，$S_{12}^i = \sin(q_1^i + q_2^i)$，$C_{123}^i = \cos(q_1^i + q_2^i + q_3^i)$ 以及 $S_{123}^i = \sin(q_1^i + q_2^i + q_3^i)$。因此，指尖坐标 $P^i(X^i, Y^i)$ 和食指方向 ϕ^i 写为

$$
\begin{cases}
X^{i} = d + L_1^{i}\cos(q_1^{i}) + L_2^{i}\cos(q_1^{i} + q_2^{i}) + L_3^{i}\cos(q_1^{i} + q_2^{i} + q_3^{i}) \\
Y^{i} = L_1^{i}\sin(q_1^{i}) + L_2^{i}\sin(q_1^{i} + q_2^{i}) + L_3^{i}\sin(q_1^{i} + q_2^{i} + q_3^{i}) \\
\phi^{i} = q_1^{i} + q_2^{i} + q_3^{i}
\end{cases} \tag{2.2.27}
$$

2.2.5 三维五指机械手

如图2.11所示，食指、中指、无名指和小指包括三个旋转关节，以便进行角运动。MCP关节位于掌骨和近端指骨之间，PIP和DIP关节分隔了指骨。拇指包括MCP和IP关节（2.1节）[1]。本书中：q_1^j，q_2^j 和 q_3^j 分别代表食指（j = i）、中指（j = m）、无名指（j = r）和小指（j = l）的第一个关节MCP^j、第二个关节PIP^j和第三个关节DIP^j的角位置；q_1^{t} 和 q_2^{t} 分别是指拇指（t）的第一个关节MCP^{t}和第二个关节IP^{t}的角位置。

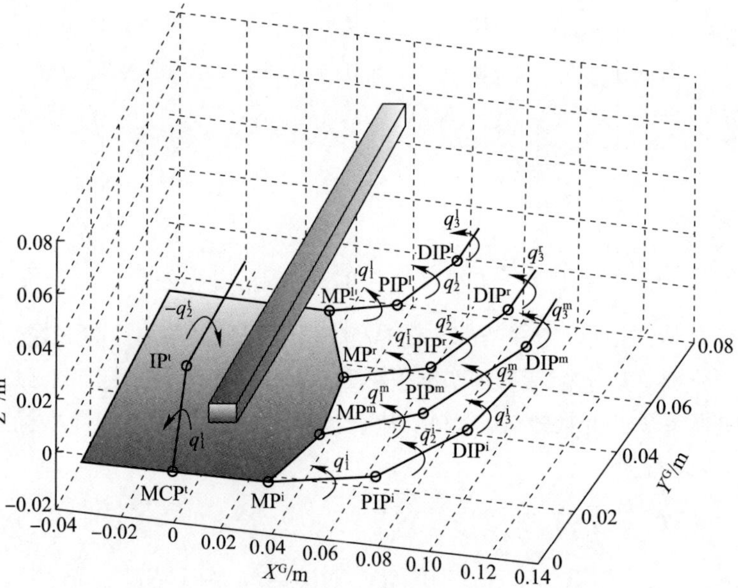

图2.11 五指机械手抓取矩形件的关节示意图

对于图2.12所示的五指机械手，X^{G}、Y^{G} 和 Z^{G} 是全局坐标系的三个轴。拇指的局部坐标系 $x^{t}-y^{t}-z^{t}$ 是通过依次将全局坐标系的 X^{G} 轴和 Y^{G} 轴分别旋转 α、β 角得到的，食指的局部坐标系 $x^{i}-y^{i}-z^{i}$ 是通过将全局坐标系的 X^{G} 轴旋转 α 角，然后转换全局坐标系的矢量 $\boldsymbol{d}^{i}$ 得到的。同样地，中指（j = m）、无名指（j = r）和小指（j = l）的局部坐标系 $x^{j}-y^{j}-z^{j}$ 是通过将全局坐标系的 X^{G} 轴旋转 α 角，然后转换全局坐标系的矢量 $\boldsymbol{d}^{j}$（j = m，r，l）得到的。

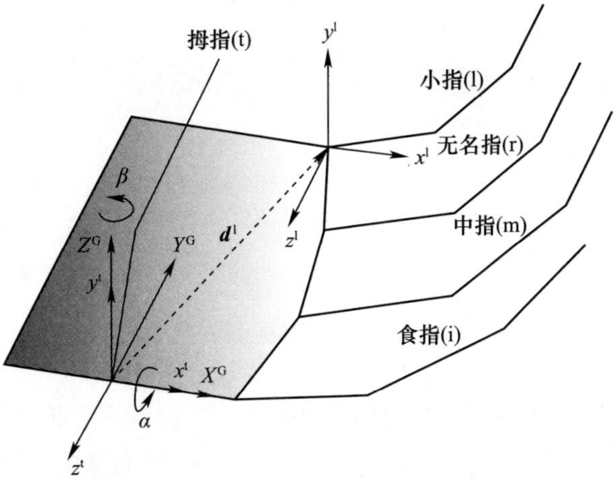

图 2.12　全局坐标系与局部坐标系的关系

$$
\begin{aligned}
{}^{\mathrm{G}}\boldsymbol{T}_{\mathrm{t}} &= R(X^{\mathrm{G}},\alpha)R(Y^{\mathrm{G}},\beta) \\
&= \begin{bmatrix} \cos\beta & 0 & \sin\beta & 0 \\ \sin\alpha\sin\beta & \cos\alpha & -\sin\alpha\cos\beta & 0 \\ -\cos\alpha\sin\beta & \sin\alpha & -\sin\alpha\cos\beta & 0 \\ 0 & 0 & 0 & 1 \end{bmatrix}
\end{aligned}
\tag{2.2.28}
$$

$$
\begin{cases}
\boldsymbol{R}(X^{\mathrm{G}},\alpha) = \begin{bmatrix} 1 & 0 & 0 & 0 \\ 0 & \cos\alpha & -\sin\alpha & 0 \\ 0 & \sin\alpha & \cos\alpha & 0 \\ 0 & 0 & 0 & 1 \end{bmatrix} \\
\boldsymbol{R}(Y^{\mathrm{G}},\beta) = \begin{bmatrix} \cos\beta & 0 & \sin\beta & 0 \\ 0 & 1 & 0 & 0 \\ -\sin\beta & 0 & \cos\beta & 0 \\ 0 & 0 & 0 & 1 \end{bmatrix}
\end{cases}
\tag{2.2.29}
$$

设 $\boldsymbol{p}^{\mathrm{G}} = [p_X^{\mathrm{G}} \quad p_Y^{\mathrm{G}} \quad p_Z^{\mathrm{G}}]'$ 和 $\boldsymbol{p}^{\mathrm{t}} = [p_X^{\mathrm{G}} \quad p_Y^{\mathrm{G}} \quad p_Z^{\mathrm{G}}]'$ 分别为全局坐标系 $X^{\mathrm{G}}-Y^{\mathrm{G}}-Z^{\mathrm{G}}$ 和拇指的局部基准坐标系 $x^{\mathrm{t}}-y^{\mathrm{t}}-z^{\mathrm{t}}$ 中任意点 P 的位置矢量。因此，$\tilde{\boldsymbol{p}}^{\mathrm{G}} = [p_X^{\mathrm{G}} \quad p_Y^{\mathrm{G}} \quad p_Z^{\mathrm{G}}]'$ 是 ${}^{\mathrm{G}}\boldsymbol{T}_{\mathrm{t}}$ 和 $\tilde{\boldsymbol{p}}^{\mathrm{t}} = [p_x^{\mathrm{t}} \quad p_y^{\mathrm{t}} \quad p_z^{\mathrm{t}} \quad 1]'$ 的乘积，即

$$
\begin{aligned}
\tilde{\boldsymbol{p}}^{\mathrm{G}} &= {}^{\mathrm{G}}\boldsymbol{T}_{\mathrm{t}}\tilde{\boldsymbol{p}}^{\mathrm{t}} \\
&= \begin{bmatrix} \cos\beta & 0 & \sin\beta & 0 \\ \sin\alpha\sin\beta & \cos\alpha & -\sin\alpha\cos\beta & 0 \\ -\cos\alpha\sin\beta & \sin\alpha & \cos\alpha\cos\beta & 0 \\ 0 & 0 & 0 & 1 \end{bmatrix} \begin{bmatrix} p_x^{\mathrm{t}} \\ p_y^{\mathrm{t}} \\ p_z^{\mathrm{t}} \\ 1 \end{bmatrix}
\end{aligned}
$$

$$
= \begin{bmatrix} \cos\beta p_x^{t} + \sin\beta p_z^{t} \\ \sin\alpha\sin\beta p_x^{t} + \cos\alpha p_y^{t} - \sin\alpha\cos\beta p_z^{t} \\ -\cos\alpha\sin\beta p_x^{t} + \sin\alpha p_y^{t} + \cos\alpha\cos\beta p_z^{t} \\ 1 \end{bmatrix} \tag{2.2.30}
$$

$\tilde{\boldsymbol{p}}^{t}$ 也由${}^{G}\boldsymbol{T}_{t}^{-1}$和 $\tilde{\boldsymbol{p}}^{G}$ 的乘积得出,即

$$
\tilde{\boldsymbol{p}}^{t} = {}^{G}\boldsymbol{T}_{t}^{-1}\tilde{\boldsymbol{p}}^{G} \tag{2.2.31}
$$

图2.12还显示了食指(i)的局部坐标系 $x^{i}-y^{i}-z^{i}$ 是通过全局坐标系 $X^{G}-Y^{G}-Z^{G}$绕 X^{G} 轴旋转 α 角,然后再沿着矢量 $\boldsymbol{d}^{i}$ 平移获得的。同样,中指($j=$m)、无名指($j=$r)和小指($j=$l)的局部坐标系 $x^{j}-y^{j}-z^{j}$ 是通过绕 X^{G} 轴旋转 α 角,然后将矢量 $\boldsymbol{d}^{j}=[d_x^{j}\quad d_y^{j}\quad d_z^{j}]'$($j=$m,r和l)相对于全局坐标系进行平移获得的。因此,将三关节手指局部坐标系 $x^{j}-y^{j}-z^{j}$($j=$m,r和l)转换为全局坐标系 $X^{G}-Y^{G}-Z^{G}$ 的齐次变换矩阵${}^{G}\boldsymbol{T}_{j}$ 表示为4个基本齐次变换 $\boldsymbol{R}(X^{G},\alpha)$、$\boldsymbol{D}(X^{G},d_x^{j})$、$\boldsymbol{D}(Y^{G},d_y^{j})$和$\boldsymbol{D}(Z^{G},d_z^{j})$的乘积。

$$
\begin{aligned}
{}^{G}\boldsymbol{T}_{j} &= \boldsymbol{R}(X^{G},\alpha)\boldsymbol{D}(X^{G},d_x^{j})\boldsymbol{D}(Y^{G},d_y^{j})\boldsymbol{D}(Z^{G},d_z^{j}) \\
&= \begin{bmatrix} 1 & 0 & 0 & d_x^{j} \\ 0 & \cos\alpha & -\sin\alpha & d_y^{j} \\ 0 & \sin\alpha & \cos\alpha & d_z^{j} \\ 0 & 0 & 0 & 1 \end{bmatrix}
\end{aligned} \tag{2.2.32}
$$

式中

$$
\begin{cases}
\boldsymbol{D}(X^{G},d_x^{j}) = \begin{bmatrix} 1 & 0 & 0 & d_x^{j} \\ 0 & 1 & 0 & 0 \\ 0 & 0 & 1 & 0 \\ 0 & 0 & 0 & 1 \end{bmatrix} \\
\boldsymbol{D}(Y^{G},d_y^{j}) = \begin{bmatrix} 1 & 0 & 0 & 0 \\ 0 & 1 & 0 & d_y^{j} \\ 0 & 0 & 1 & 0 \\ 0 & 0 & 0 & 1 \end{bmatrix} \\
\boldsymbol{D}(Z^{G},d_z^{j}) = \begin{bmatrix} 1 & 0 & 0 & 0 \\ 0 & 1 & 0 & 0 \\ 0 & 0 & 1 & d_z^{j} \\ 0 & 0 & 0 & 1 \end{bmatrix}
\end{cases} \tag{2.2.33}
$$

设 $\boldsymbol{p}^j = [p_x^j \quad p_y^j \quad p_z^j]'$为三关节手指局部基准坐标系中任意点 P 的位置矢量。则 $\boldsymbol{p}^G$ 可由 ${}^G\boldsymbol{T}_j$ 和 $\tilde{\boldsymbol{p}}^j = [p_x^j \quad p_y^j \quad p_z^j \quad 1]'$的乘积计算,即

$$
\begin{aligned}
\tilde{\boldsymbol{p}}^G &= {}^G\boldsymbol{T}_j \tilde{\boldsymbol{p}}^j \\
&= \begin{bmatrix} 1 & 0 & 0 & d_x^j \\ 0 & \cos\alpha & -\sin\alpha & d_y^j \\ 0 & \sin\alpha & \cos\alpha & d_z^j \\ 0 & 0 & 0 & 1 \end{bmatrix} \begin{bmatrix} p_x^j \\ p_y^j \\ p_z^j \\ 1 \end{bmatrix} \\
&= \begin{bmatrix} p_x^j + d_x^j \\ \cos\alpha p_y^j - \sin\alpha p_z^j + d_y^j \\ \sin\alpha p_y^j + \cos\alpha p_z^j + d_z^j \\ 1 \end{bmatrix}
\end{aligned} \tag{2.2.34}
$$

$\tilde{\boldsymbol{p}}^j$ 也由 ${}^G\boldsymbol{T}_j^{-1}$和 $\tilde{\boldsymbol{p}}^G$ 的乘积计算,即

$$
\tilde{\boldsymbol{p}}^j = {}^G\boldsymbol{T}_j^{-1} \tilde{\boldsymbol{p}}^G \tag{2.2.35}
$$

2.3 反向运动

理想轨迹(2.5 节)通常定义在笛卡尔坐标系中,并且轨迹控制在关节空间更容易实现。因此,有必要将笛卡尔坐标系转化到关节空间中[6-7,9-10]。使用反向动力学,每一根手指的关节角度需要从已知的指尖位置(关节空间)获取。关节的角速度和角加速度通过几何雅可比矩阵从指尖(末端执行器)的线速度、角速度和加速度获得。

2.3.1 双关节拇指

每一根手指的关节角度推导如下。根据正向动力学[6-7,9-10](2.2 节),拇指(t)的指尖坐标 $P^t(X^t, Y^t)$描述如下:

$$
\begin{cases} X^t = L_1^t \cos(q_1^t) + L_2^t \cos(q_1^t + q_2^t) \\ Y^t = L_1^t \sin(q_1^t) + L_2^t \sin(q_1^t + q_2^t) \end{cases} \tag{2.3.1}
$$

式中:L_1^t 和 L_2^t 分别是拇指的关节 1 和关节 2 的长度;q_1^t 和 q_2^t 是拇指的关节 1 和关节 2 的角度。式(2.3.1)的平方和为

$$
\begin{aligned}
X^{t2} + Y^{t2} &= L_1^{t2}\cos^2(q_1^t) + L_2^{t2}\cos^2(q_1^t + q_2^t) + 2L_1^{t2}L_2^{t2}\cos(q_1^t)\cos(q_1^t + q_2^t) + \\
&\quad L_1^{t2}\sin^2(q_1^t) + L_2^{t2}\sin^2(q_1^t + q_2^t) + 2L_1^{t2}L_2^{t2}\sin(q_1^t)\sin(q_1^t + q_2^t) \\
&= L_1^{t2} + L_2^{t2} + 2L_1^{t2}L_2^{t2}\cos(q_2^t)
\end{aligned} \tag{2.3.2}
$$

重新整理公式(2.3.2)可得

$$\cos(q_2^t)=\frac{X^{t2}+Y^{t2}-L_1^{t2}-L_2^{t2}}{2L_1^tL_2^t} \tag{2.3.3}$$

根据上肘关节的构造,关节 2 的角度 q_2^t 由以下公式获得:

$$q_2^t=-\cos^{-1}\frac{X^{t2}+Y^{t2}-L_1^{t2}-L_2^{t2}}{2L_1^tL_2^t} \tag{2.3.4}$$

注意,在这本书中,所有逆时针旋转的角度定义为正值。当按照上肘关节的构造时,角度 q_2^t 是顺时针的,所以 q_2^t 的符号为负。

图2.13 为双关节拇指(上肘)的几何学说明[2,13]。基于几何学,我们可以得到如下的两个正切关系:

$$\begin{cases}\tan(\alpha^t)=\dfrac{Y^t}{X^t}\\ \tan(\beta^t)=\dfrac{L_2^t\sin(q_2^t)}{L_1^t+L_2^t\cos(q_2^t)}\end{cases} \tag{2.3.5}$$

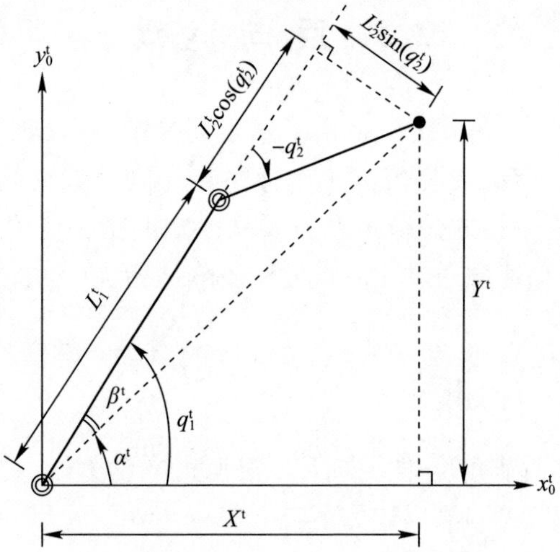

图2.13 双关节拇指(上肘)的几何学说明

根据式(2.3.5),可得 α^t 和 β^t:

$$\begin{cases}\alpha^t=\tan^{-1}\left(\dfrac{Y^t}{X^t}\right)\\ \beta^t=\tan^{-1}\left[\dfrac{L_2^t\sin(q_2^t)}{L_1^t+L_2^t\cos(q_2^t)}\right]\end{cases} \tag{2.3.6}$$

然后，通过式(2.3.6)可得关节1的角度 q_1^t 为

$$\begin{aligned} q_1^t &= \alpha^t + \beta^t \\ &= \tan^{-1}\left(\frac{Y^t}{X^t}\right) - \tan^{-1}\left[\frac{L_2^t \sin(q_2^t)}{L_1^t + L_2^t \cos(q_2^t)}\right] \end{aligned} \tag{2.3.7}$$

因此，反向动力学给出了关节角度的表达式(式(2.3.4)和式(2.3.7))。

2.3.2 三关节食指

在抓取东西的过程中，Kamper 等[14]和 Luo 等[15]研究了人类力量控制策略和关节运动轨迹形式。Kamper 等[14]的研究说明，最符合人类食指指尖运动轨迹的数据曲线为对数形式，该数据由10个人(年龄在21~32岁之间)演示的20次抓取试验得到。Zollo 等[16]的研究表明，极坐标(r,θ)表示为

$$r = [1.3394(L_1^i + L_2^i + L_3^i) - 23.255]\exp(-0.062\theta) \tag{2.3.8}$$

由于食指、中指、无名指和小指被认为是三关节手指(2.2节)，因此食指的反向动力学模型代表了所有三关节手指。基于2.2.4节的正向动力学，食指(i)在指尖笛卡尔坐标系(X^i, Y^i)中的对数表达形式用三个关节变量 q_1^i、q_2^i 和 q_3^i 表示为

$$\begin{cases} X^i = d + L_1^i\cos(q_1^i) + L_2^i\cos(q_1^i + q_2^i) + L_3^i\cos(q_1^i + q_2^i + q_3^i) \\ Y^i = L_1^i\sin(q_1^i) + L_2^i\sin(q_1^i + q_2^i) + L_3^i\sin(q_1^i + q_2^i + q_3^i) \end{cases} \tag{2.3.9}$$

这里基于实验数据，文献[14-16]使用 $q_3^i = 0.7q_2^i$ 来解决图2.10中平面 $X^G - Y^G$ 的冗余问题。将 $q_3^i = 0.7q_2^i$ 代入式(2.3.9)中，则有

$$\begin{cases} X^i = d + L_1^i\cos(q_1^i) + L_2^i\cos(q_1^i + q_2^i) + L_3^i\cos(q_1^i + 1.7q_2^i) \\ Y^i = L_1^i\sin(q_1^i) + L_2^i\sin(q_1^i + q_2^i) + L_3^i\sin(q_1^i + 1.7q_2^i) \end{cases} \tag{2.3.10}$$

非线性函数(2.3.10)有多种求解方法，如 Newton - Raphson[9]法，Wen 等提出的 HPSONN[17]法，Chen 等提出的 GA[12,18-19]法，Chen 等提出的 ANFIS 法[12,18-19]。现在，通过设置关于变量 q_1^i 和 q_2^i 的两个函数 $f_1^i(q_1^i, q_2^i)$ 和 $f_2^i(q_1^i, q_2^i)$，可以得到

$$\begin{cases} f_1^i(q_1^i, q_2^i) = d + L_1^i\cos(q_1^i) + L_2^i\cos(q_1^i + q_2^i) + L_3^i\cos(q_1^i + 1.7q_2^i) - X^i \\ f_2^i(q_1^i, q_2^i) = L_1^i\sin(q_1^i) + L_2^i\sin(q_1^i + q_2^i) + L_3^i\sin(q_1^i + 1.7q_2^i) - Y^i \end{cases} \tag{2.3.11}$$

这样，问题变为关于变量 q_1^i 和 q_2^i 的两个函数 $f_1^i(q_1^i,q_2^i)$ 和 $f_2^i(q_1^i,q_2^i)$ 的最优值(最小值)问题。用遗传算法搜寻 q_1^{i*} 和 q_2^{i*} 使函数 $f_1^i(q_1^i,q_2^i)$ 和 $f_2^i(q_1^i,q_2^i)$ 的值接近0，此时 q_1^{i*}、q_2^{i*} 和 q_3^{i*} ($=0.7q_2^{i*}$) 为指尖坐标系(X^i,Y^i)下关节角度的解。或者，反向动力学问题可以使用 ANFIS 方法[20]，此时模糊神经系统的输入为笛卡尔坐标系，输出为关节坐标系。在仿真[12-13,18-19]过程中，我们发现 GA 法给出了一个更好的结果(误差 $\approx 10^{-7}$)，但是需要更多运算时间，ANFIS 法相较于 GA 法给出了一个较好的结果(误差 $\approx 10^{-4}$)，但需要较少的运算时间，ANFIS 法和 GA 法将分别在 4.3 节和 4.5 节详细介绍。只要求解出上述的角位置，那么食指的所有关节角速度和角加速度都可以通过 2.4 节的差动运动学计算出来。相似地，中指、无名指和小指的所有关节角速度和角加速度也都可以计算出来。

2.3.3 指尖工作区域

2.3.3.1 双关节拇指与三关节食指

当双关节拇指和三关节食指做扩展(屈伸)运动时，指尖运动范围被 MCP、PIP 和 DCP 的关节角度限制。换句话说，当手指的所有关节实现所有可能的运动时，指尖可到达的空间称为“可达工作空间”。每个关节的最大关节角度存在个体差异。根据反向动力学，图 2.14 展示了双关节拇指的工作空间。第一个和第二个关节角度位置(关节角度)分别限制在了[0°,90°]和[−80°,0°]之间。*区域代表双关节拇指的工作空间(到目前为止我们还没有考虑方向)。

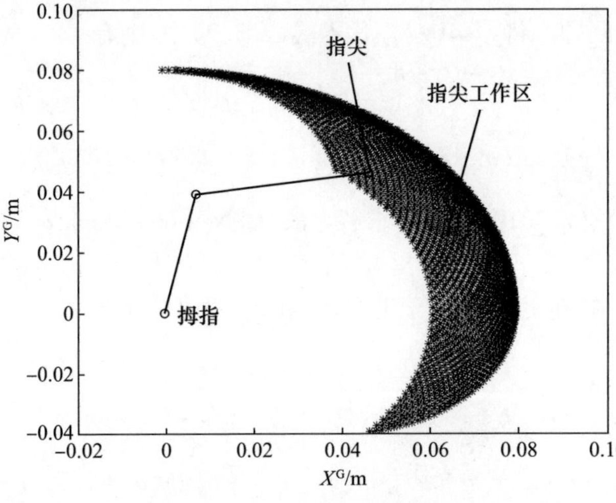

图 2.14　双关节拇指的工作空间

相似地，三关节食指的工作空间如图 2.15 所示，第一个（MCP）、第二个（PIP）和第三个（DIP）食指关节角度分别限制在[0°,90°]、[0°,110°]和[0°,80°]之间[21]，食指指尖的可达空间为 * 区域。图 2.16 用方形物体将图 2.14 和图 2.15 组合起来[12]。因此，下部和上部分别展示了拇指和食指指尖的可达位置。其中，重叠区域代表拇指和食指均可到达的区域。拇指的两个关节长度均为 0.040m，食指的三个关节长度分别为 0.040m、0.040m 和 0.030m[22]。

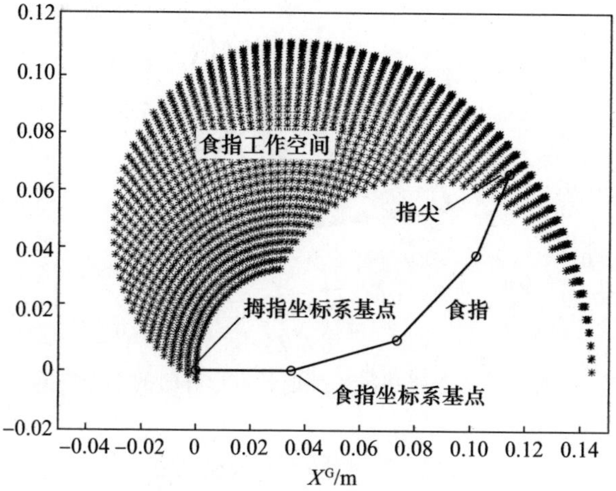

图 2.15　三关节食指的工作空间

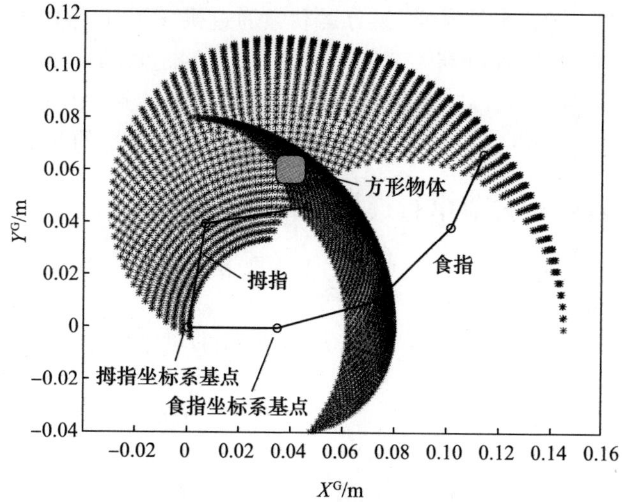

图 2.16　抓取方形物体时拇指和食指的工作空间

2.3.3.2 五指机械手

图2.17展示了五指机械手(14自由度)抓取长方形杆时的工作空间。另外3根手指(中指、无名指、小指)的第一个、第二个和第三个关节角度被分别限制在[0°,90°]、[0°,110°]和[0°,80°]之间[21]。图中,灰色区域分别表示拇指、食指、中指、无名指和小指指尖的可达位置。注意到X^G轴设置为不同的长度,这是为了展示清楚每根手指的三维活动范围[18]。

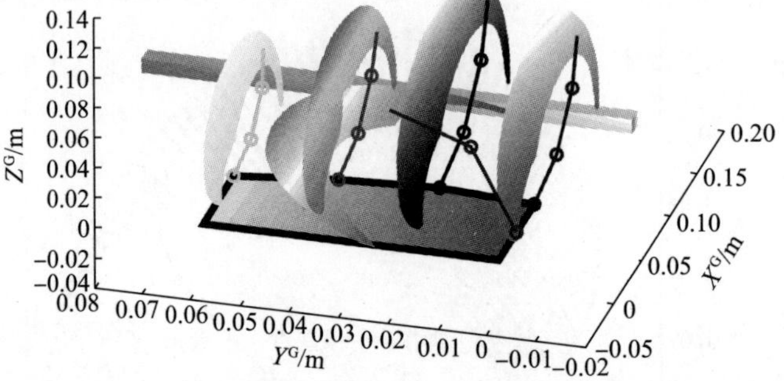

图2.17 五指机械手抓取长方形杆时的工作空间

2.4 微分运动学

2.2节(正向动力学)和2.3节(反向动力学)推导了指尖位置和关节角位置的关系。这节将介绍微分运动学,微分运动学通过机械手几何雅可比矩阵建立指尖的线速度、角速度、加速度与关节的角速度、加速度之间的关系。在计算几何雅可比矩阵之前,需要先回顾一下旋转矩阵的性质和刚体运动学。下面将计算多旋转关节串行连接的平面机械手、双关节拇指、三关节食指的几何雅可比矩阵。

2.4.1 多旋转关节串行连接的二维机械手

根据2.2节的正向运动学,式(2.2.18)是多旋转关节串行连接的平面机械手正向运动学公式。

$$
\begin{aligned}
{}^0\boldsymbol{T}_n(q) &= {}^0\boldsymbol{T}_1(\boldsymbol{q})\,{}^1\boldsymbol{T}_2(\boldsymbol{q})\cdots{}^{i-1}\boldsymbol{T}_i(\boldsymbol{q})\cdots{}^{n-1}\boldsymbol{T}_n(\boldsymbol{q}) \\
&= \begin{bmatrix} C_{12\cdots n} & -S_{12\cdots n} & 0 & \sum\limits_{i=1}^{n} L_i C_{12\cdots i} \\ S_{12\cdots n} & C_{12\cdots n} & 0 & \sum\limits_{i=1}^{n} L_i S_{12\cdots j} \\ 0 & 0 & 1 & 0 \\ 0 & 0 & 0 & 1 \end{bmatrix}
\end{aligned}
$$

$$= \begin{bmatrix} {}^0\boldsymbol{R}_n(\boldsymbol{q}) & {}^0\boldsymbol{P}_n(\boldsymbol{q}) \\ 0 & 1 \end{bmatrix} \tag{2.4.1}$$

式中：$\boldsymbol{q}=[q_1 \quad q_2 \quad \cdots \quad q_n]'$为 n 个关节的角度位置矢量；${}^0\boldsymbol{T}_n(\boldsymbol{q})$为齐次变换矩阵，将末端执行器转移到坐标系中；${}^0\boldsymbol{R}_n(\boldsymbol{q})$和${}^0\boldsymbol{P}_n(\boldsymbol{q})$分别为旋转矩阵（方向）和转移矢量（位置），将末端执行器转移到坐标系中。使用了符号 $C_{12\cdots n}=\cos(q_1+q_2+\cdots+q_n)$和 $S_{12\cdots n}=\sin(q_1+q_2+\cdots+q_n)$。

末端执行器在坐标系下的线速度${}^0\boldsymbol{P}_n(\boldsymbol{q})$和角速度${}^0\boldsymbol{\omega}_n$与关节角速度 $\dot{\boldsymbol{q}}$ 成线性关系，它们之间的线性关系可以用差动运动学方程表示，方程如下：

$${}^0\boldsymbol{V}_n=\boldsymbol{J}(\boldsymbol{q})\dot{\boldsymbol{q}} \tag{2.4.2}$$

式中：${}^0\boldsymbol{V}_n=[{}^0\dot{\boldsymbol{P}}_n \quad {}^0\boldsymbol{\omega}_n]$为$(6\times1)$末端执行器的速度矢量；$\boldsymbol{J}(q)=[\boldsymbol{J}_P(\boldsymbol{q}) \quad \boldsymbol{J}_O(\boldsymbol{q})]'$为$(6\times n)$系列多旋转关节串行连接的平面机械手的几何雅可比矩阵，其中位置雅可比矩阵 $\boldsymbol{J}_P(q)$和方向雅可比矩阵 $\boldsymbol{J}_O(\boldsymbol{q})$均为$(3\times n)$矩阵，它们分别将关节角速度 $\dot{\boldsymbol{q}}$ 转变为末端执行器的线速度${}^0\dot{\boldsymbol{P}}_n$和角速度${}^0\boldsymbol{\omega}_n$。在计算几何雅可比矩阵 $\boldsymbol{J}(q)$之前，需要先回顾一下旋转矩阵和刚体运动学。

2.4.1.1 旋转矩阵特性

式(2.4.1)中，对转移矢量${}^0\boldsymbol{P}_n(\boldsymbol{q})$（位置）和旋转矩阵${}^0\boldsymbol{R}_n(\boldsymbol{q})$（方向）而言，末端执行器的姿态是关节角度位置矢量 $\boldsymbol{q}$ 的函数。利用微分运动学的目的是描述末端执行器的线速度${}^0\dot{\boldsymbol{P}}_n(\boldsymbol{q})$和角速度${}^0\boldsymbol{\omega}_n$，所以有必要考虑含时间参数的旋转矩阵对于时间 t 的一阶导数性质。

因为时间依赖矩阵 $\boldsymbol{R}(t)$为正交矩阵，所以有如下性质：

$$\boldsymbol{R}(t)\boldsymbol{R}'(t)=\boldsymbol{I} \tag{2.4.3}$$

对式(2.4.3)进行微分，得到方程

$$\dot{\boldsymbol{R}}(t)\boldsymbol{R}'(t)+\boldsymbol{R}(t)\dot{\boldsymbol{R}}'(t)=\boldsymbol{0} \tag{2.4.4}$$

设 $\boldsymbol{S}(t)=\dot{\boldsymbol{R}}(t)\boldsymbol{R}'(t)$，式(2.4.4)可改写为

$$\boldsymbol{S}(t)+\boldsymbol{S}'(t)=\boldsymbol{0} \tag{2.4.5}$$

$\boldsymbol{S}(t)$为(3×3)的反对称矩阵。式(2.4.4)两边同时乘 $\boldsymbol{R}(t)$可得

$$\dot{\boldsymbol{R}}(t)\boldsymbol{R}'(t)\boldsymbol{R}(t)=-\boldsymbol{R}(t)\dot{\boldsymbol{R}}'(t)\boldsymbol{R}(t) \tag{2.4.6}$$

将式(2.4.3)和式(2.4.5)代入式(2.4.6)可得

$$\dot{\boldsymbol{R}}(t) = -[\boldsymbol{R}(t)\dot{\boldsymbol{R}}'(t)]\boldsymbol{R}(t)$$
$$= \boldsymbol{S}(t)\boldsymbol{R}(t) \tag{2.4.7}$$

式(2.4.7)给出了 $\boldsymbol{R}(t)$ 关于时间的一阶导数。考虑矢量 $\boldsymbol{p}(t) = \boldsymbol{R}(t)\boldsymbol{p}'$，其中 $\boldsymbol{p}'$ 为常数矢量。$\boldsymbol{R}(t)$ 对于时间 t 的一阶导数为

$$\dot{\boldsymbol{p}}(t) = \dot{\boldsymbol{R}}(t)\boldsymbol{p}' \tag{2.4.8}$$

将式(2.4.7)代入式(2.4.8)，式(2.4.8)可改写为

$$\dot{\boldsymbol{p}}(t) = \boldsymbol{S}(t)\boldsymbol{R}(t)\boldsymbol{p}' \tag{2.4.9}$$

如果矢量 $\boldsymbol{\omega}(t)$ 代表参考坐标系在时间 t 下经过 $\boldsymbol{R}(t)$ 变换后坐标系中的角速度，那么有如下关系：

$$\dot{\boldsymbol{p}}(t) = \boldsymbol{\omega}(t) \times \boldsymbol{R}(t)\boldsymbol{p}' \tag{2.4.10}$$

因此，比较式(2.4.9)和式(2.4.10)，我们可以知道矢量 $\boldsymbol{S}(t)$ 介于 $\boldsymbol{\omega}(t) = [\omega_x \quad \omega_y \quad \omega_z]'$ 和 $\boldsymbol{R}(t)\boldsymbol{p}'$ 之间。$\boldsymbol{S}(t)$ 为

$$\boldsymbol{S}(t) = \begin{bmatrix} 0 & -\omega_z & \omega_y \\ \omega_z & 0 & -\omega_x \\ -\omega_y & \omega_x & 0 \end{bmatrix}$$
$$= \boldsymbol{S}(\boldsymbol{\omega}(t)) \tag{2.4.11}$$

因此，式(2.4.7)可改写为

$$\dot{\boldsymbol{R}}(t) = \boldsymbol{S}(\boldsymbol{\omega}(t))\boldsymbol{R}(t) \tag{2.4.12}$$

除此之外，下面的等式参见文献[10]。

$$\boldsymbol{R}(t)\boldsymbol{S}(\boldsymbol{\omega}(t))\boldsymbol{R}'(t) = \boldsymbol{S}(\boldsymbol{R}(t)\boldsymbol{\omega}(t)) \tag{2.4.13}$$

2.4.1.2 刚体运动学

在图 2.3 中，式(2.2.6)表明了转移矢量 ${}^0\boldsymbol{o}_1$ 和旋转矩阵 ${}^0\boldsymbol{R}_1$ 从坐标系 1 到坐标系 0 的坐标系转换关系。

$$\boldsymbol{p}^0 = {}^0\boldsymbol{o}_1 + {}^0\boldsymbol{R}_1\boldsymbol{p}^1 \tag{2.4.14}$$

式(2.4.14)关于时间的一阶导数为

$$\dot{\boldsymbol{p}}^0 = {}^0\dot{\boldsymbol{o}}_1 + {}^0\boldsymbol{R}_1\dot{\boldsymbol{p}}^1 + {}^0\dot{\boldsymbol{R}}_1\boldsymbol{p}^1 \tag{2.4.15}$$

将式(2.4.12)代入式(2.4.15)，式(2.4.15)改写为

$$\dot{\boldsymbol{p}}^0 = {}^0\dot{\boldsymbol{o}}_1 + {}^0\boldsymbol{R}_1\dot{\boldsymbol{p}}^1 + \boldsymbol{S}({}^0\boldsymbol{\omega}_1){}^0\boldsymbol{R}_1\boldsymbol{p}^1$$

$$= {}^0\dot{\boldsymbol{o}}_1 + {}^0\boldsymbol{R}_1\dot{\boldsymbol{p}}^1 + {}^0\boldsymbol{\omega}_1 \times {}^0\boldsymbol{p}_1 \tag{2.4.16}$$

式中

$${}^0\boldsymbol{p}_1 = {}^0\boldsymbol{R}_1\boldsymbol{p}^1$$

图 2.18 给出了关节 i 的矢量表示,关节 i 连接了节点 i 和节点$(i+1)$。坐标系$(i-1)$固连于关节$(i-1)$,原点为 o_{i-1} 且在节点 i 上。坐标系 i 固连于关节 i,原点为 o_i 且在节点$(i+1)$上。${}^0\boldsymbol{p}_{i-1}$ 和 ${}^0\boldsymbol{p}_i$ 分别为在坐标系$(i-1)$和坐标系 i 下的位置矢量。根据式(2.4.14),从坐标系$(i-1)$到坐标系 i 的坐标系转换公式为

$${}^0\boldsymbol{p}_i = {}^0\boldsymbol{p}_{i-1} + {}^0\boldsymbol{R}_{i-1} + {}^{i-1}\boldsymbol{p}_i^{i-1} \tag{2.4.17}$$

式中:${}^{i-1}\boldsymbol{p}_i^{i-1}$ 为坐标系$(i-1)$(右上标)中坐标原点 o_i(右下标)相对于坐标原点 o_{i-1}(左上标)的位置矢量。式(2.4.17)关于时间的导数如下:

$${}^0\dot{\boldsymbol{p}}_i = {}^0\dot{\boldsymbol{p}}_{i-1} + {}^0\boldsymbol{R}_{i-1}{}^{i-1}\dot{\boldsymbol{p}}_i^{i-1} + {}^0\dot{\boldsymbol{R}}_{i-1}^{i-1}\boldsymbol{p}_i^{i-1}$$

$$= {}^0\dot{\boldsymbol{p}}_{i-1} + {}^{i-1}\dot{\boldsymbol{p}}_i + {}^0\boldsymbol{\omega}_{i-1} \times {}^{i-1}\boldsymbol{p}_i \tag{2.4.18}$$

式(2.4.18)说明了关节 i 的线速度 ${}^0\dot{\boldsymbol{p}}_i$ 是关于关节$(i-1)$的转移速度$({}^0\dot{\boldsymbol{p}}_{i-1} + {}^{i-1}\dot{\boldsymbol{p}}_i)$和旋转速度$({}^0\boldsymbol{\omega}_{i-1} \times {}^{i-1}\boldsymbol{p}_i)$的函数。

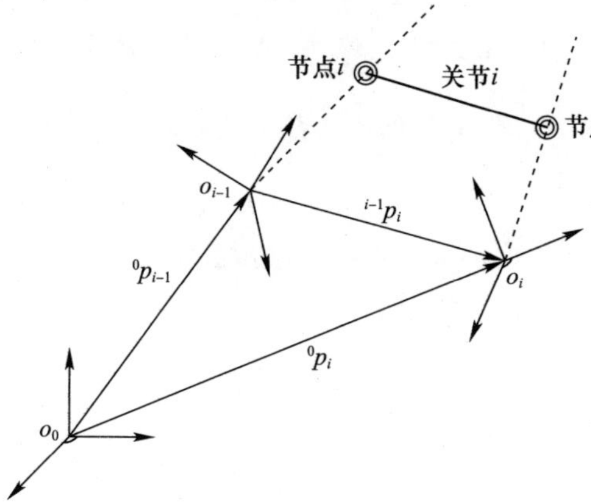

图 2.18　关节 i 的矢量表示

下面将给出关节角速度的推导过程。旋转结构为

$${}^0\boldsymbol{R}_i = {}^0\dot{\boldsymbol{R}}_{i-1}^{i-1}\boldsymbol{R}_i \tag{2.4.19}$$

式(2.4.19)关于时间的导数为

$$^{0}\dot{\boldsymbol{R}}_{i}={}^{0}\dot{\boldsymbol{R}}_{i-1}^{i-1}\boldsymbol{R}_{i}+{}^{0}\dot{\boldsymbol{R}}_{i-1}^{i-1}\dot{\boldsymbol{R}}_{i} \tag{2.4.20}$$

将式(2.4.12)代入式(2.4.20)可得

$$\begin{aligned}\boldsymbol{S}({}^{0}\boldsymbol{\omega}_{i}){}^{0}\boldsymbol{R}_{i}&=\boldsymbol{S}({}^{0}\omega_{i-1}){}^{0}\boldsymbol{R}_{i-1}^{i-1}+{}^{0}\boldsymbol{R}_{i-1}\boldsymbol{S}({}^{i-1}\boldsymbol{\omega}_{i}^{i-1}){}^{i-1}\boldsymbol{R}_{i}\\&=\boldsymbol{S}({}^{0}\omega_{i-1}){}^{0}\boldsymbol{R}_{i-1}^{i-1}\boldsymbol{R}_{i}+{}^{0}\boldsymbol{R}_{i-1}\boldsymbol{S}({}^{i-1}\boldsymbol{\omega}_{i}^{i-1}){}^{0}\boldsymbol{R}'^{0}_{i-1}\boldsymbol{R}_{i-1}^{i-1}\boldsymbol{R}_{i}\\&=\boldsymbol{S}({}^{0}\omega_{i-1}){}^{0}\boldsymbol{R}_{i}+\boldsymbol{S}({}^{0}\boldsymbol{R}_{i-1}^{i-1}\boldsymbol{\omega}_{i}^{i-1}){}^{0}\boldsymbol{R}_{i}\\&=\boldsymbol{S}({}^{0}\omega_{i-1}){}^{0}\boldsymbol{R}_{i}+\boldsymbol{S}({}^{i-1}\boldsymbol{\omega}_{i}){}^{0}\boldsymbol{R}_{i}\end{aligned} \tag{2.4.21}$$

由此得到以下结果:

$$^{0}\boldsymbol{\omega}_{i}={}^{0}\boldsymbol{\omega}_{i-1}+{}^{i-1}\boldsymbol{\omega}_{i} \tag{2.4.22}$$

式(2.4.22)说明,关节 i 的角速度 $^{0}\boldsymbol{\omega}_{i}$ 是关节$(i-1)$的角速度 $^{0}\boldsymbol{\omega}_{i-1}$ 和关节 i 关于关节$(i-1)$的角速度 $^{i-1}\boldsymbol{\omega}_{i}$ 的函数。

对于旋转关节,坐标系 i 关于坐标系$(i-1)$的旋转是由关节 i 的移动引发。因此式(2.4.18)右边的第二部分可以改写为

$$^{i-1}\dot{\boldsymbol{p}}_{i}={}^{i-1}\boldsymbol{\omega}_{i}\times{}^{i-1}\boldsymbol{p}_{i} \tag{2.4.23}$$

将式(2.4.23)代入式(2.4.18)可得

$$\begin{aligned}{}^{0}\dot{\boldsymbol{p}}_{i}&={}^{0}\dot{\boldsymbol{p}}_{i-1}+{}^{i-1}\boldsymbol{\omega}_{i}\times{}^{i-1}\boldsymbol{p}_{i}+{}^{0}\boldsymbol{\omega}_{i-1}\times{}^{i-1}\boldsymbol{p}_{i}\\&={}^{0}\dot{\boldsymbol{p}}_{i-1}+({}^{i-1}\boldsymbol{\omega}_{i}+{}^{0}\boldsymbol{\omega}_{i-1})\times{}^{i-1}\boldsymbol{p}_{i}\\&={}^{0}\dot{\boldsymbol{p}}_{i-1}+{}^{0}\boldsymbol{\omega}_{i}\times{}^{i-1}\boldsymbol{p}_{i}\end{aligned} \tag{2.4.24}$$

式(2.4.22)改写为

$$^{0}\boldsymbol{\omega}_{i}={}^{0}\boldsymbol{\omega}_{i-1}+\dot{q}_{i}\boldsymbol{z}_{i-1} \tag{2.4.25}$$

式中:$\boldsymbol{z}_{i-1}$为关节 i 所在轴的单位矢量。

将式(2.4.24)和式(2.4.25)代入式(2.4.2)可得

$$\begin{bmatrix}{}^{0}\dot{\boldsymbol{p}}_{n}\\{}^{0}\boldsymbol{\omega}_{n}\end{bmatrix}=\begin{bmatrix}\boldsymbol{J}_{P}(q)\\\boldsymbol{J}_{O}(q)\end{bmatrix}\dot{q}=\begin{bmatrix}\boldsymbol{J}_{P1}&\boldsymbol{J}_{P2}&\cdots&\boldsymbol{J}_{Pn}\\\boldsymbol{J}_{O1}&\boldsymbol{J}_{O2}&\cdots&\boldsymbol{J}_{On}\end{bmatrix}\begin{bmatrix}\dot{q}_{1}\\\dot{q}_{2}\\\vdots\\\dot{q}_{n}\end{bmatrix}$$

$$= \begin{bmatrix} \sum_{i=1}^{n} \boldsymbol{J}_{pi}\dot{q}_i \\ \sum_{i=1}^{n} \boldsymbol{J}_{Oi}\dot{q}_i \end{bmatrix} = \begin{bmatrix} \sum_{i=1}^{n} \frac{\partial^0 \boldsymbol{p}_n}{\partial q_i}\dot{q}_i \\ \sum_{i=1}^{n} {}^{i-1}\boldsymbol{\omega}_i \end{bmatrix} = \begin{bmatrix} \sum_{i=1}^{n} {}^{i-1}\boldsymbol{\omega}_i \times {}^{i-1}\boldsymbol{p}_n \\ \sum_{i=1}^{n} \dot{q}_i \boldsymbol{z}_{i-1} \end{bmatrix}$$

$$= \begin{bmatrix} \sum_{i=1}^{n} \dot{q}_i \boldsymbol{z}_{i-1} \times ({}^0\boldsymbol{p}_n - {}^0\boldsymbol{p}_{i-1}) \\ \sum_{i=1}^{n} \dot{q}_i \boldsymbol{z}_{i-1} \end{bmatrix} \tag{2.4.26}$$

因此,几何雅可比矩阵$\boldsymbol{J}(\boldsymbol{q})$为

$$\boldsymbol{J}(q) = \begin{bmatrix} \boldsymbol{J}_P(\boldsymbol{q}) \\ \boldsymbol{J}_O(\boldsymbol{q}) \end{bmatrix} = \begin{bmatrix} \boldsymbol{J}_{P1} & \boldsymbol{J}_{P2} & \cdots & \boldsymbol{J}_{Pn} \\ \boldsymbol{J}_{O1} & \boldsymbol{J}_{O2} & \cdots & \boldsymbol{J}_{On} \end{bmatrix} \tag{2.4.27}$$

式中:$[\boldsymbol{J}_{Pi} \quad \boldsymbol{J}_{Oi}]'(i=1,2,\cdots,n)$由下式计算得到,即

$$\begin{bmatrix} \boldsymbol{J}_{Pi} \\ \boldsymbol{J}_{Oi} \end{bmatrix} = \begin{bmatrix} \boldsymbol{z}_{i-1} \times ({}^0\boldsymbol{P}_n - {}^0\boldsymbol{p}_{i-1}) \\ \boldsymbol{z}_{i-1} \end{bmatrix} \tag{2.4.28}$$

关节i所在轴的单位矢量z_{i-1}是从旋转矩阵的第三列获得的,表达式为

$$z_{i-1} = {}^0\boldsymbol{R}_1(q_1)\,{}^1\boldsymbol{R}_2(q_2)\cdots{}^{i-2}\boldsymbol{R}_{i-1}(q_{i-1})\boldsymbol{z}_0 \tag{2.4.29}$$

式中:$\boldsymbol{z}_0 = [0 \quad 0 \quad 1]' = z_j (j=1,2,\cdots,n)$。

末端执行器的位置矢量${}^0\boldsymbol{P}_n$由齐次变换矩阵${}^0\boldsymbol{T}_n$第四列的前三个值获得,表达式为

$$\begin{aligned} {}^0\tilde{\boldsymbol{p}}_n &= {}^0\boldsymbol{T}_n^0\tilde{\boldsymbol{p}}_n \\ &= {}^0\boldsymbol{T}_1(q_1)\,{}^1\boldsymbol{T}_2(q_2)\cdots{}^{n-1}\boldsymbol{T}_n(q_n)\,{}^0\tilde{\boldsymbol{p}}_n \end{aligned} \tag{2.4.30}$$

式中:${}^0\tilde{\boldsymbol{p}}_n = [{}^0\boldsymbol{p}_n \quad 1]'$,${}^0\tilde{\boldsymbol{p}}_0 = [0 \quad 0 \quad 0 \quad 1]'$。因此${}^0\boldsymbol{P}_n$可由下式计算得到:

$${}^0\boldsymbol{P}_n = \begin{bmatrix} \sum_{i=1}^{n} L_i C_{12\cdots i} \\ \sum_{i=1}^{n} L_i S_{12\cdots i} \\ 0 \end{bmatrix} \tag{2.4.31}$$

式中：$C_{12\cdots j}=\cos(q_1+q_2+\cdots+q_j)$；$S_{12\cdots j}=\sin(q_1+q_2+\cdots+q_j)$。

相似地，关节 i 的位置矢量 ${}^0\boldsymbol{p}_{i-1}$ 由齐次变换矩阵 ${}^0\boldsymbol{T}_{i-1}$ 第四列的前三个值获得，表达式为

$$\begin{aligned}{}^0\tilde{\boldsymbol{p}}_{i-1}&={}^0\boldsymbol{T}_{i-1}^0\tilde{\boldsymbol{p}}_0\\&={}^0\boldsymbol{T}_1(q_1){}^1\boldsymbol{T}_2(q_2)\cdots{}^{i-2}\boldsymbol{T}_{i-1}(q_{i-1}){}^0\tilde{\boldsymbol{p}}_0\end{aligned}\tag{2.4.32}$$

式中：${}^0\tilde{\boldsymbol{p}}_{i-1}=[{}^0p_{i-1}\quad 1]'$。因此，${}^0\tilde{\boldsymbol{p}}_{i-1}(i=2,3,\cdots,n)$ 可由下式计算得到：

$${}^0\boldsymbol{p}_{i-1}=\begin{bmatrix}\sum\limits_{j=1}^{i-1}L_jC_{12\cdots(i-1)}\\\sum\limits_{j=1}^{i-1}L_jS_{12\cdots(i-1)}\\0\end{bmatrix}\tag{2.4.33}$$

式中：$C_{12\cdots(i-1)}=\cos(q_1+q_2+\cdots+q_{i-1})$；$S_{12\cdots(i-1)}=\sin(q_1+q_2+\cdots+q_{i-1})$。将式(2.4.29)、式(2.4.31)和式(2.4.33)代入式(2.4.28)，式(2.4.27)中的几何雅可比矩阵 $\boldsymbol{J}(\boldsymbol{q})$ 可改写为

$$\begin{aligned}\boldsymbol{J}(\boldsymbol{q})&=\begin{bmatrix}\boldsymbol{J}_{P1}&\boldsymbol{J}_{P2}&\cdots&\boldsymbol{J}_{Pn}\\\boldsymbol{J}_{O1}&\boldsymbol{J}_{O2}&\cdots&\boldsymbol{J}_{On}\end{bmatrix}\\&=\begin{bmatrix}\boldsymbol{z}_0\times({}^0\boldsymbol{P}_n-{}^0p_0)&\boldsymbol{z}_1\times({}^0\boldsymbol{P}_n-{}^0\boldsymbol{p}_1)&\cdots&\boldsymbol{z}_n\times({}^0\boldsymbol{P}_n-{}^0\boldsymbol{p}_{n-1})\\\boldsymbol{z}_0&\boldsymbol{z}_1&\cdots&\boldsymbol{z}_n\end{bmatrix}\\&=\begin{bmatrix}-\sum\limits_{i=1}^{n}L_iS_{12\cdots n}&-\sum\limits_{i=2}^{n}L_iS_{12\cdots n}&\cdots&-\sum\limits_{i=n}^{n}L_iS_{12\cdots n}\\\sum\limits_{i=1}^{n}L_iC_{12\cdots n}&\sum\limits_{i=2}^{n}L_iC_{12\cdots n}&\cdots&\sum\limits_{i=n}^{n}L_iC_{12\cdots n}\\0&0&\cdots&0\\0&0&\cdots&0\\0&0&\cdots&0\\1&1&\cdots&1\end{bmatrix}\end{aligned}\tag{2.4.34}$$

几何雅可比矩阵 $\boldsymbol{J}(\boldsymbol{q})$ 的 J_{ij} 项可改写为

$$J_{ij}=\begin{cases}-\sum_{j=1}^{n}L_j\sin\left(\sum_{k=1}^{j}q_k\right), i=1,j\in[1,6]\\ \sum_{j=1}^{n}L_j\cos\left(\sum_{k=1}^{j}q_k\right), i=2,j\in[1,6]\\ 0, i\in[3,5], j\in[1,6]\\ 1, i=6, j\in[1,6]\end{cases}$$

接着,角速度 $\dot{\boldsymbol{q}}$ 可以改写为

$$\dot{\boldsymbol{q}}=\boldsymbol{J}^{-1}(\boldsymbol{q})^{0}\boldsymbol{V}_n \tag{2.4.35}$$

相似地,角加速度为

$$\ddot{\boldsymbol{q}}=\boldsymbol{J}^{-1}(\boldsymbol{q})[{}^{0}\boldsymbol{A}_n-\dot{\boldsymbol{J}}(\boldsymbol{q})\dot{\boldsymbol{q}}] \tag{2.4.36}$$

式中:${}^{0}\boldsymbol{A}_n=[{}^{0}\ddot{P}_n\quad {}^{0}\alpha_n]'$为$(6\times1)$末端执行器的加速度矢量,包括线加速度${}^{0}\ddot{P}_n$和角加速度${}^{0}\alpha_n$。差动雅可比矩阵 $\dot{\boldsymbol{J}}(q)$可改写为文献[5]中:

$$\dot{J}_{ij}=\begin{cases}-\sum_{j=1}^{n}L_j\cos\left(\sum_{k=1}^{j}q_k\right)\left(\sum_{k=1}^{j}\dot{q}_k\right), i=1,j\in[1,6]\\ -\sum_{j=1}^{n}L_j\sin\left(\sum_{k=1}^{j}q_k\right)\left(\sum_{k=1}^{j}\dot{q}_k\right), i=1,j\in[1,6]\\ 0, i\in[3,6], j\in[1,6]\end{cases}$$

2.4.2 双关节拇指

如果只考虑指尖的线速度,那么可以通过式(2.3.1)的一阶导数和二阶导数得到关节的角速度和角加速度。指尖相应的线速度 $\dot{p}^{\mathrm{t}}(\dot{X}^{\mathrm{t}},\dot{Y}^{\mathrm{t}})=\mathrm{d}(\dot{X}^{\mathrm{t}},\dot{Y}^{\mathrm{t}})/\mathrm{d}t$ 可由下式获得:

$$\begin{bmatrix}\dot{X}^{\mathrm{t}}\\ \dot{Y}^{\mathrm{t}}\end{bmatrix}=\begin{bmatrix}-L_1^{\mathrm{t}}\sin(q_1^{\mathrm{t}})-L_2^{\mathrm{t}}\sin(q_1^{\mathrm{t}}+q_2^{\mathrm{t}}) & -L_2^{\mathrm{t}}\sin(q_1^{\mathrm{t}}+q_2^{\mathrm{t}})\\ L_1^{\mathrm{t}}\cos(q_1^{\mathrm{t}})+L_2^{\mathrm{t}}\cos(q_1^{\mathrm{t}}+q_2^{\mathrm{t}}) & L_2^{\mathrm{t}}\cos(q_1^{\mathrm{t}}+q_2^{\mathrm{t}})\end{bmatrix}$$

或者表示为矩阵形式:

$$\dot{\boldsymbol{P}}^{\mathrm{t}}=\boldsymbol{J}_P^{\mathrm{t}}(\boldsymbol{q}^{\mathrm{t}})\dot{\boldsymbol{q}}^{\mathrm{t}} \tag{2.4.37}$$

矩阵 $\dot{\boldsymbol{P}}^{t}$、$\dot{\boldsymbol{q}}^{t}$、$\boldsymbol{q}^{t}$、$\boldsymbol{J}_P^{t}(\boldsymbol{q}^{t})$分别为

$$\begin{cases}\dot{\boldsymbol{P}}^{t}=\begin{bmatrix}\dot{X}^{t}\\ \dot{Y}^{t}\end{bmatrix},\dot{\boldsymbol{q}}^{t}=\begin{bmatrix}\dot{q}_1^{t}\\ \dot{q}_2^{t}\end{bmatrix},\boldsymbol{q}^{t}=\begin{bmatrix}q_1^{t}\\ q_2^{t}\end{bmatrix}\\ \boldsymbol{J}_P^{t}(q^{t})=\begin{bmatrix}-L_1^{t}\sin(q_1^{t})-L_2^{t}\sin(q_1^{t}+q_2^{t}) & -L_2^{t}\sin(q_1^{t}+q_2^{t})\\ L_1^{t}\cos(q_1^{t})+L_2^{t}\cos(q_1^{t}+q_2^{t}) & L_2^{t}\cos(q_1^{t}+q_2^{t})\end{bmatrix}\end{cases} \tag{2.4.38}$$

矩阵 $\boldsymbol{J}_P^{t}(\boldsymbol{q}^{t})$为拇指的几何雅可比子矩阵。完整的几何雅可比矩阵，如式(2.3.34)，是(6×2)的矩阵，它的最后三行分别表示每一个关节的角速度。我们只需要考虑几何雅可比矩阵的平面位置 $P^{t}(\dot{X}^{t},\dot{Y}^{t})$，而不需要考虑方向。

关节 1 和关节 2 的角速度 $\dot{q}_1^{t}$和 $\dot{q}_2^{t}$为

$$\dot{\boldsymbol{q}}^{t}=\boldsymbol{J}_P^{t}(\boldsymbol{q}^{t})^{-1}\dot{\boldsymbol{P}}^{t} \tag{2.4.39}$$

类似地，关节 1 和关节 2 的角加速度 $\ddot{q}_1^{t}$ 和 $\ddot{q}_2^{t}$ 为

$$\ddot{\boldsymbol{q}}^{t}=\boldsymbol{J}_P^{t}(\boldsymbol{q}^{t})^{-1}\left[\ddot{\boldsymbol{P}}^{t}-\frac{\mathrm{d}\boldsymbol{J}_P^{t}(\boldsymbol{q}^{t})}{\mathrm{d}t}\dot{\boldsymbol{q}}^{t}\right] \tag{2.4.40}$$

式中：$\ddot{\boldsymbol{P}}^{t}$ 为指尖的线加速度。$\ddot{\boldsymbol{P}}^{t}$、$\ddot{\boldsymbol{q}}^{t}$ 和 $\mathrm{d}\boldsymbol{J}_P^{t}(\boldsymbol{q}^{t})/\mathrm{d}t$ 为

$$\ddot{\boldsymbol{P}}^{t}=\begin{bmatrix}\ddot{X}^{t}\\ \ddot{Y}^{t}\end{bmatrix},\ddot{\boldsymbol{q}}^{t}=\begin{bmatrix}\ddot{q}_1^{t}\\ \ddot{q}_2^{t}\end{bmatrix} \tag{2.4.41}$$

$$\frac{\mathrm{d}\boldsymbol{J}_P^{t}(\boldsymbol{q}^{t})}{\mathrm{d}t}=\begin{bmatrix}-L_1^{t}\cos(q_1^{t})\dot{q}_1^{t}-L_2^{t}\cos(q_1^{t}+q_2^{t})(\dot{q}_1^{t}+\dot{q}_2^{t}) & -L_2^{t}\cos(q_1^{t}+q_2^{t})(\dot{q}_1^{t}+\dot{q}_2^{t})\\ -L_1^{t}\sin(q_1^{t})\dot{q}_1^{t}-L_2^{t}\sin(q_1^{t}+q_2^{t})(\dot{q}_1^{t}+\dot{q}_2^{t}) & -L_2^{t}\sin(q_1^{t}+q_2^{t})(\dot{q}_1^{t}+\dot{q}_2^{t})\end{bmatrix} \tag{2.4.42}$$

2.4.3 三关节食指

与双关节拇指相似，通过计算食指位置的一阶导数，可以得到线速度 $\mathrm{d}(X^{i},Y^{i})/\mathrm{d}t$ 为

$$\dot{\boldsymbol{P}}^{t}=\boldsymbol{J}_P^{t}(\boldsymbol{q}^{t})\dot{\boldsymbol{q}}^{t} \tag{2.4.43}$$

矩阵 $\dot{\boldsymbol{P}}^{\mathrm{i}}$、$\dot{\boldsymbol{q}}^{\mathrm{i}}$、$\boldsymbol{q}^{\mathrm{i}}$ 和 $\boldsymbol{J}^{\mathrm{i}}(\boldsymbol{q})$分别为

$$\dot{\boldsymbol{P}}^{\mathrm{i}}=\begin{bmatrix}\dot{X}^{\mathrm{i}}\\ \dot{Y}^{\mathrm{i}}\end{bmatrix},\ \dot{\boldsymbol{q}}^{\mathrm{i}}=\begin{bmatrix}\dot{q}_1^{\mathrm{i}}\\ \dot{q}_2^{\mathrm{i}}\end{bmatrix},\boldsymbol{q}^{\mathrm{i}}=\begin{bmatrix}q_1^{\mathrm{i}}\\ q_2^{\mathrm{i}}\end{bmatrix},\boldsymbol{J}_P^{\mathrm{i}}(\boldsymbol{q}^{\mathrm{i}})=\begin{bmatrix}J_{11}^{\mathrm{i}}(q^{\mathrm{i}}) & J_{12}^{\mathrm{i}}(q^{\mathrm{i}})\\ J_{21}^{\mathrm{i}}(q^{\mathrm{i}}) & J_{22}^{\mathrm{i}}(q^{\mathrm{i}})\end{bmatrix} \tag{2.4.44}$$

式中

$$\begin{cases}J_{11}^{\mathrm{i}}=-L_1^{\mathrm{i}}\sin(q_1^{\mathrm{i}})-L_2^{\mathrm{i}}\sin(q_1^{\mathrm{i}}+q_2^{\mathrm{i}})-L_3^{\mathrm{i}}\sin(q_1^{\mathrm{i}}+1.7q_2^{\mathrm{i}})\\ J_{12}^{\mathrm{i}}=-L_2^{\mathrm{i}}\sin(q_1^{\mathrm{i}}+q_2^{\mathrm{i}})-1.7L_3^{\mathrm{i}}\sin(q_1^{\mathrm{i}}+1.7q_2^{\mathrm{i}})\\ J_{21}^{\mathrm{i}}=L_1^{\mathrm{i}}\cos(q_1^{\mathrm{i}})+L_2^{\mathrm{i}}\cos(q_1^{\mathrm{i}}+q_2^{\mathrm{i}})+L_3^{\mathrm{i}}\cos(q_1^{\mathrm{i}}+1.7q_2^{\mathrm{i}})\\ J_{22}^{\mathrm{i}}=L_2^{\mathrm{i}}\cos(q_1^{\mathrm{i}}+q_2^{\mathrm{i}})+1.7L_3^{\mathrm{i}}\cos(q_1^{\mathrm{i}}+1.7q_2^{\mathrm{i}})\end{cases} \tag{2.4.45}$$

相应地,关节 1 和关节 2 的角速度 $\dot{q}_1^{\mathrm{i}}$和 $\dot{q}_2^{\mathrm{i}}$为

$$\dot{\boldsymbol{q}}^{\mathrm{i}}=\boldsymbol{J}_P^{\mathrm{i}}(\boldsymbol{q}^{\mathrm{i}})^{-1}\dot{\boldsymbol{P}}^{\mathrm{i}} \tag{2.4.46}$$

类似地,关节 3 的角速度 $\dot{q}_3^{\mathrm{i}}$ 为

$$\dot{q}_3^{\mathrm{i}}=0.7\,\dot{q}_2^{\mathrm{i}} \tag{2.4.47}$$

用相似的方法,关节 1 和关节 2 的角加速度 $\ddot{q}_1^{\mathrm{i}}$ 和 $\ddot{q}_2^{\mathrm{i}}$ 为

$$\ddot{\boldsymbol{q}}^{\mathrm{i}}=\boldsymbol{J}_P^{\mathrm{i}}(\boldsymbol{q}^{\mathrm{i}})^{-1}\left[\ddot{\boldsymbol{P}}^{\mathrm{i}}-\frac{\mathrm{d}\boldsymbol{J}_P^{\mathrm{i}}(\boldsymbol{q}^{\mathrm{i}})}{\mathrm{d}t}\dot{\boldsymbol{q}}^{\mathrm{i}}\right] \tag{2.4.48}$$

式中:$\ddot{\boldsymbol{P}}^{\mathrm{i}}$ 为指尖的线加速度。$\ddot{\boldsymbol{P}}^{\mathrm{i}}$、$\ddot{\boldsymbol{q}}^{\mathrm{i}}$ 和 $\mathrm{d}\boldsymbol{J}_P^{\mathrm{i}}(\boldsymbol{q}^{\mathrm{i}})/\mathrm{d}t$ 分别为

$$\ddot{\boldsymbol{P}}^{\mathrm{i}}=\begin{bmatrix}\ddot{X}^{\mathrm{i}}\\ \ddot{Y}^{\mathrm{i}}\end{bmatrix},\ddot{\boldsymbol{q}}^{\mathrm{i}}=\begin{bmatrix}\ddot{q}_1^{\mathrm{t}}\\ \ddot{q}_2^{\mathrm{t}}\end{bmatrix},\frac{\mathrm{d}\boldsymbol{J}_P^{\mathrm{i}}(q^{\mathrm{i}})}{\mathrm{d}t}=\begin{bmatrix}\dot{J}_{11}^{\mathrm{i}}(q^{\mathrm{i}}) & \dot{J}_{12}^{\mathrm{i}}(q^{\mathrm{i}})\\ \dot{J}_{21}^{\mathrm{i}}(q^{\mathrm{i}}) & \dot{J}_{22}^{\mathrm{i}}(q^{\mathrm{i}})\end{bmatrix} \tag{2.4.49}$$

式中

$$\begin{cases}\dot{J}_{11}^{\mathrm{i}}(q^{\mathrm{i}})=-L_1^{\mathrm{i}}\cos(q_1^{\mathrm{i}})\,\dot{q}_1^{\mathrm{i}}-L_2^{\mathrm{i}}\cos(q_1^{\mathrm{i}}+q_2^{\mathrm{i}})(\dot{q}_1^{\mathrm{i}}+\dot{q}_2^{\mathrm{i}})-L_3^{\mathrm{i}}\cos(q_1^{\mathrm{i}}+1.7q_2^{\mathrm{i}})(\dot{q}_1^{\mathrm{i}}+1.7\,\dot{q}_2^{\mathrm{i}})\\ \dot{J}_{12}^{\mathrm{i}}(q^{\mathrm{i}})=-L_2^{\mathrm{i}}\cos(q_1^{\mathrm{i}}+q_2^{\mathrm{i}})(\dot{q}_1^{\mathrm{i}}+\dot{q}_2^{\mathrm{i}})-1.7L_3^{\mathrm{i}}\cos(q_1^{\mathrm{i}}+1.7q_2^{\mathrm{i}})(\dot{q}_1^{\mathrm{i}}+1.7\,\dot{q}_2^{\mathrm{i}})\\ \dot{J}_{21}^{\mathrm{i}}(q^{\mathrm{i}})=-L_1^{\mathrm{i}}\sin(q_1^{\mathrm{i}})\,\dot{q}_1^{\mathrm{i}}-L_2^{\mathrm{i}}\sin(q_1^{\mathrm{i}}+q_2^{\mathrm{i}})(\dot{q}_1^{\mathrm{i}}+\dot{q}_2^{\mathrm{i}})-L_3^{\mathrm{i}}\sin(q_1^{\mathrm{i}}+1.7q_2^{\mathrm{i}})(\dot{q}_1^{\mathrm{i}}+1.7\,\dot{q}_2^{\mathrm{i}})\\ \dot{J}_{22}^{\mathrm{i}}(q^{\mathrm{i}})=-L_2^{\mathrm{i}}\sin(q_1^{\mathrm{i}}+q_2^{\mathrm{i}})(\dot{q}_1^{\mathrm{i}}+\dot{q}_2^{\mathrm{i}})-1.7L_3^{\mathrm{i}}\sin(q_1^{\mathrm{i}}+1.7q_2^{\mathrm{i}})(\dot{q}_1^{\mathrm{i}}+1.7\,\dot{q}_2^{\mathrm{i}})\end{cases} \tag{2.4.50}$$

关节 3 的角加速度 $\ddot{q}_3^{\mathrm{i}}$ 为

$$\ddot{q}_3^{\mathrm{i}} = 0.7\ddot{q}_2^{\mathrm{i}} \tag{2.4.51}$$

2.5 轨迹规划

在机械手执行特定的移动任务之前,得知最理想的路径很重要。轨迹规划包括期望路径的生成或者参考输入,与一些特殊情况有关,如避开障碍物,期望路径不能超过执行器的电压和扭矩极限。最简单的机械手移动为点对点移动。2.5.1 节和 2.5.2 节分别介绍由最初值和最终值描述的包括位置和速度的三次多项式时间序列和贝塞尔曲线函数产生的期望路径。

2.5.1 三次多项式轨迹规划

多项式函数定义为评估一个多项式的函数,通常关于变量 t 的多项式函数 $P(t)$ 为

$$P(t) = \sum_{i=0}^{n} A_i t^i \tag{2.5.1}$$

式中:n 为非负整数;$A_i(i=0,1,2,\cdots,n)$ 为常系数。

为了产生一个光滑的轨迹,指尖空间的三次多项式如图 2.19 所示。

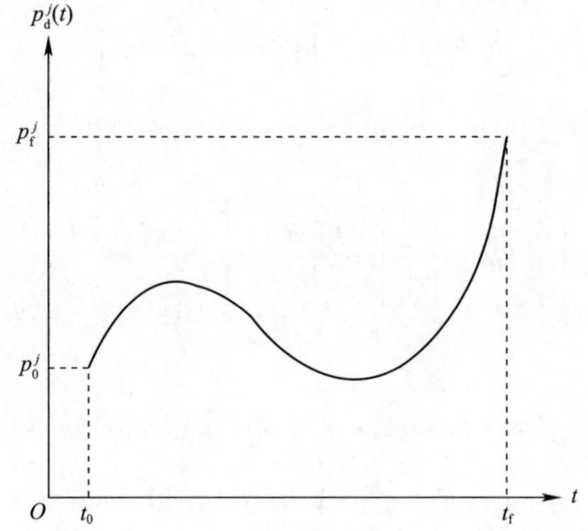

图 2.19　指尖空间轨迹

指尖位置(p)、速度(v)和加速度(a)随时间的变化关系见文献[9,10,23],即有

$$\begin{cases} p_{\mathrm{d}}^{j}(t) = A_0 + A_1 t + A_2 t^2 + A_3 t^3 \\ v_{\mathrm{d}}^{j}(t) = A_1 + 2A_2 t + 3A_3 t^2 \\ a_{\mathrm{d}}^{j}(t) = 2A_2 + 6A_3 t \end{cases} \tag{2.5.2}$$

式中：$A_0 \sim A_3$ 为不确定常数；上标 j 表示手指的序号，如 j 为 t、i、m、r 和 l，分别代表拇指、食指、中指、无名指和小指。

假如需要在 t_0 和 t_{f} 处满足以下 4 个限定条件：

$$\begin{cases} p_{\mathrm{d}}^{j}(t_0) = p_0^j \\ p_{\mathrm{d}}^{j}(t_0) = p_0^j \\ p_{\mathrm{d}}^{j}(t_{\mathrm{f}}) = p_{\mathrm{f}}^j \\ v_{\mathrm{d}}^{j}(t_{\mathrm{f}}) = v_{\mathrm{f}}^j \end{cases} \tag{2.5.3}$$

式中：t_0 和 t_{f} 分别为初始时间和终止时间；p_0^j 和 p_{f}^j 分别为手指 j 指尖位置的初始值和终值，基于 t_0 和 t_{f} 处的 4 个限定条件，式(2.5.2)可改写为

$$\begin{cases} p_{\mathrm{d}}^{j}(t_0) = A_0 + A_1 t_0 + A_2 t_0^2 + A_3 t_0^3 = p_0^j \\ v_{\mathrm{d}}^{j}(t_0) = A_1 + 2A_2 t_0 + 3A_3 t_0^2 = v_0^j \\ p_{\mathrm{d}}^{j}(t_{\mathrm{f}}) = A_0 + A_1 t_{\mathrm{f}} + A_2 t_{\mathrm{f}}^2 + A_3 t_{\mathrm{f}}^3 = p_{\mathrm{f}}^j \\ v_{\mathrm{d}}^{j}(t_{\mathrm{f}}) = A_1 + 2A_2 t_{\mathrm{f}} + 3A_3 t_{\mathrm{f}}^2 = v_{\mathrm{f}}^j \end{cases} \tag{2.5.4}$$

用矩阵形式表示，式(2.5.4)可改写为

$$\boldsymbol{T}^{P}\boldsymbol{A} = \boldsymbol{P} \tag{2.5.5}$$

式中：矩阵 $\boldsymbol{T}^{P}$、$\boldsymbol{A}$ 和 $\boldsymbol{P}$ 分别为

$$\boldsymbol{T}^{P} = \begin{bmatrix} 1 & t_0 & t_0^2 & t_0^3 \\ 0 & 1 & 2t_0 & 3t_0^2 \\ 1 & t_{\mathrm{f}} & t_{\mathrm{f}}^2 & t_{\mathrm{f}}^3 \\ 0 & 1 & 2t_{\mathrm{f}} & 3t_{\mathrm{f}}^2 \end{bmatrix}$$

$$\boldsymbol{A} = [A_0 \quad A_1 \quad A_2 \quad A_3]'$$

$$\boldsymbol{P} = [p_0^j \quad v_0^j \quad p_{\mathrm{f}}^j \quad v_{\mathrm{f}}^j]' \tag{2.5.6}$$

因此，$A_0 \sim A_3$ 这 4 个不确定常数可由下式计算：

$$\boldsymbol{A} = \boldsymbol{T}^{P-1}\boldsymbol{P} \tag{2.5.7}$$

2.5.2 三次贝塞尔曲线轨迹规划

贝塞尔曲线在计算机图像中被广泛地用于模拟光滑曲线，更高维度的贝塞尔曲线被称为贝塞尔曲面。贝塞尔曲线由法国工程师贝塞尔于 1962 年提出，用来设计汽车车身。1959 年，Paul de Casteljau 发展了该曲线，他使用了数值收敛

的方法来评估贝塞尔曲线[24]。

贝塞尔曲线函数 $B(t)$ 通常表示为

$$B(t) = \sum_{i=0}^{n} B_i b_i^n(t), t \in [0,1] \tag{2.5.8}$$

式中：n 为非负整数；$B_i(i=0,1,2,\cdots,n)$ 为常系数；$b_i^n(t)$ 为

$$\begin{aligned} b_i^n(t) &= C_i^n t^i (1-t)^{n-i} = \binom{n}{i} t^i (1-t)^{n-i} \\ &= \frac{n!}{i!\ (n-i)!} t^i (1-t)^{n-i} \end{aligned} \tag{2.5.9}$$

类似地，指尖位置（p）、速度（v）和加速度（a）的三次贝赛尔曲线随时间的变化关系如下：

$$\left\{\begin{aligned} p_{\mathrm{d}}^j(\hat{t}) &= B_0(1-\hat{t})^3 + 3B_1\hat{t}(1-\hat{t})^2 + 3B_2\hat{t}^2(1-\hat{t}) + B_3\hat{t}^3 \\ &= B_0(1-3\hat{t}+3\hat{t}^2-\hat{t}^3) + 3B_1(\hat{t}-2\hat{t}^2+\hat{t}^3) + 3B_2(\hat{t}^2-\hat{t}^3) + B_3\hat{t}^3 \\ v_{\mathrm{d}}^j(\hat{t}) &= [B_0(-3+6\hat{t}-3\hat{t}^2) + 3B_1(1-4\hat{t}+3\hat{t}^2) + 3B_2(2\hat{t}-3\hat{t}^2) + 3B_3\hat{t}^2]\ \dot{\hat{t}} \\ a_{\mathrm{d}}^j(\hat{t}) &= [6B_0(1-\hat{t}) + 6B_1(-2+3\hat{t})^2 + 6B_2(1-3\hat{t}) + 6B_3]\ \dot{\hat{t}}^2 \end{aligned}\right. \tag{2.5.10}$$

式中：$B_0 \sim B_3$ 为不确定常数；上标 j 为手指的序号，如 j 为 t、i、m、r 和 l，分别代表拇指、食指、中指、无名指和小指。转换时间 $\hat{t}$ 和其一阶导数 $\dot{\hat{t}}$ 分别为

$$\hat{t} = \frac{t-t_0}{t_{\mathrm{f}}-t_0} \in [0,1] \tag{2.5.11}$$

$$\dot{\hat{t}} = \frac{\mathrm{d}\hat{t}}{\mathrm{d}t} = \frac{1}{t_{\mathrm{f}}-t_0} \tag{2.5.12}$$

式中：t 为实际时间；t_0 和 t_{f} 分别为实际初始时间和真实终止时间，在实际时间 t_0 和 t_{f} 处，应该满足以下 4 个限定条件：

$$\left\{\begin{aligned} p_{\mathrm{d}}^j(\hat{t}=0) &= p_0^j \\ v_{\mathrm{d}}^j(\hat{t}=0) &= v_0^j \\ p_{\mathrm{d}}^j(\hat{t}=1) &= p_{\mathrm{f}}^j \\ v_{\mathrm{d}}^j(\hat{t}=1) &= v_{\mathrm{f}}^j \end{aligned}\right. \tag{2.5.13}$$

式中：p_0^j 和 p_{f}^j 分别为手指 j 的初始和终值指尖位置；v_0^j 和 v_{f}^j 分别为手指 j 的初始和终值指尖速度。基于 t_0 和 t_{f} 处的 4 个限定条件，式（2.5.10）可改写为

$$
\begin{cases}
p_{\mathrm{d}}^{j}(\hat{t}=0)=B_0=p_0^{j}\\
v_{\mathrm{d}}^{j}(\hat{t}=0)=\left(\dfrac{-3}{t_{\mathrm{f}}-t_0}\right)B_0+\left(\dfrac{3}{t_{\mathrm{f}}-t_0}\right)B_1=v_0^{j}\\
p_{\mathrm{d}}^{j}(\hat{t}=1)=B_3=p_{\mathrm{f}}^{j}\\
v_{\mathrm{d}}^{j}(\hat{t}=1)=\left(\dfrac{-3}{t_{\mathrm{f}}-t_0}\right)B_2+\left(\dfrac{3}{t_{\mathrm{f}}-t_0}\right)B_3=v_{\mathrm{f}}^{j}
\end{cases}
\tag{2.5.14}
$$

用矩阵形式表示,式(2.5.14)可改写为

$$
\boldsymbol{T}^{B}\boldsymbol{B}=\boldsymbol{P}
\tag{2.5.15}
$$

式中:矩阵 $\boldsymbol{T}^{B}$、$\boldsymbol{B}$ 和 $\boldsymbol{P}$ 分别为

$$
\begin{cases}
\boldsymbol{T}^{B}=\begin{bmatrix}
1 & 0 & 0 & 0\\
\dfrac{-3}{t_{\mathrm{f}}-t_0} & \dfrac{3}{t_{\mathrm{f}}-t_0} & 0 & 0\\
0 & 0 & 0 & 1\\
0 & 0 & \dfrac{-3}{t_{\mathrm{f}}-t_0} & \dfrac{3}{t_{\mathrm{f}}-t_0}
\end{bmatrix}\\
\boldsymbol{B}=[B_0\quad B_1\quad B_2\quad B_3]'\\
\boldsymbol{P}=[p_0^{j}\quad v_0^{j}\quad p_{\mathrm{f}}^{j}\quad v_{\mathrm{f}}^{j}]'
\end{cases}
\tag{2.5.16}
$$

因此,$B_0 \sim B_3$ 这 4 个不确定常数可由下式计算:

$$
\boldsymbol{B}=\boldsymbol{T}^{B-1}\boldsymbol{P}
\tag{2.5.17}
$$

2.5.3 轨迹路径仿真结果

图 2.20 ~ 图 2.22 分别展示了三次多项式和三次贝赛尔曲线函数的指尖轨迹位置、线速度和线加速度。初始位置和终端位置分别为 0.030m 和 0.065m,跟踪时间为 20s。从结果看,三次多项式和三次贝赛尔曲线函数并没有明显区别。因此,在所有的控制策略研究中将会使用多项式函数。在插值超过两个点时使用贝赛尔曲线函数,因为控制点可以更容易和直观地进行移动以获得使用者更喜欢的曲线。

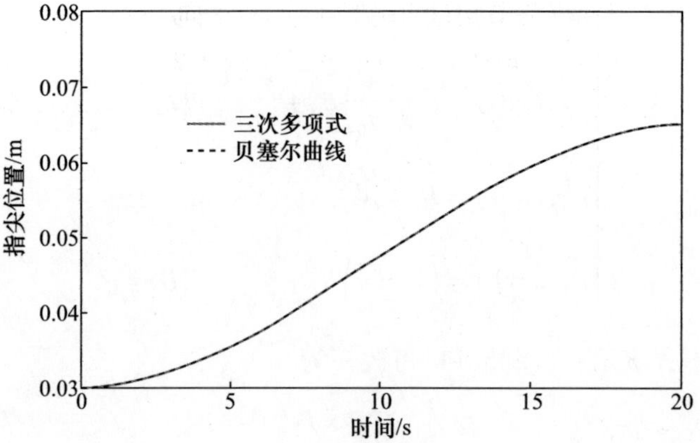

图 2.20　指尖的轨迹位置

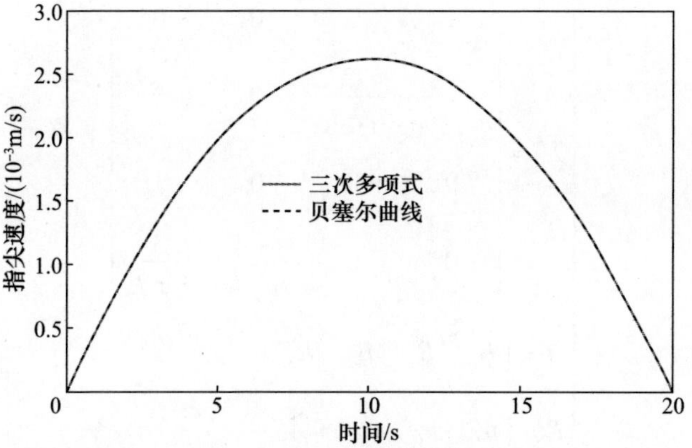

图 2.21　指尖的轨迹线速度

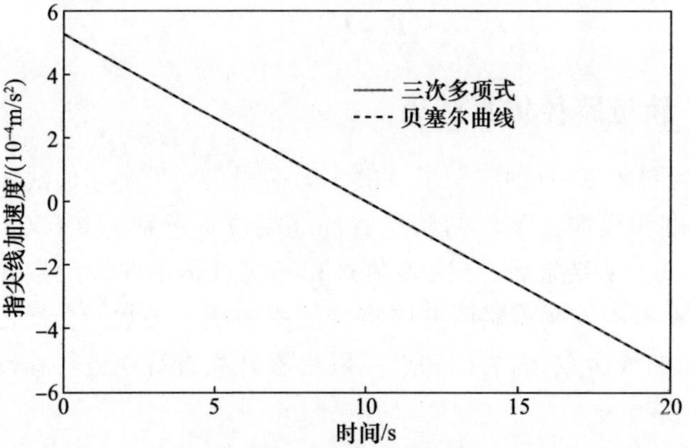

图 2.22　指尖的轨迹线加速度

图 2.23 展示了拇指关节角度和指尖位置的正向动力学(图(a))以及拇指关节位置和关节角度的反向动力学(图(b))。图 2.24 展示了食指关节角度和尖端位置的正向动力学(图(a))以及食指关节位置和关节角度的反向动力学(图(b)),使用的是 ANFIS 法(见 4.3 节),通过使用 GA 法(见 4.5 节)也获得了相似的结果。在仿真期间[12],发现 GA 法给出了一个更好的结果(error ≈ 10^{-7}),但是代价为消耗更多的运算时间,ANFIS 法给出了一个较好的结果(error ≈ 10^{-4}),同时比 GA 法需要更少的运算时间。

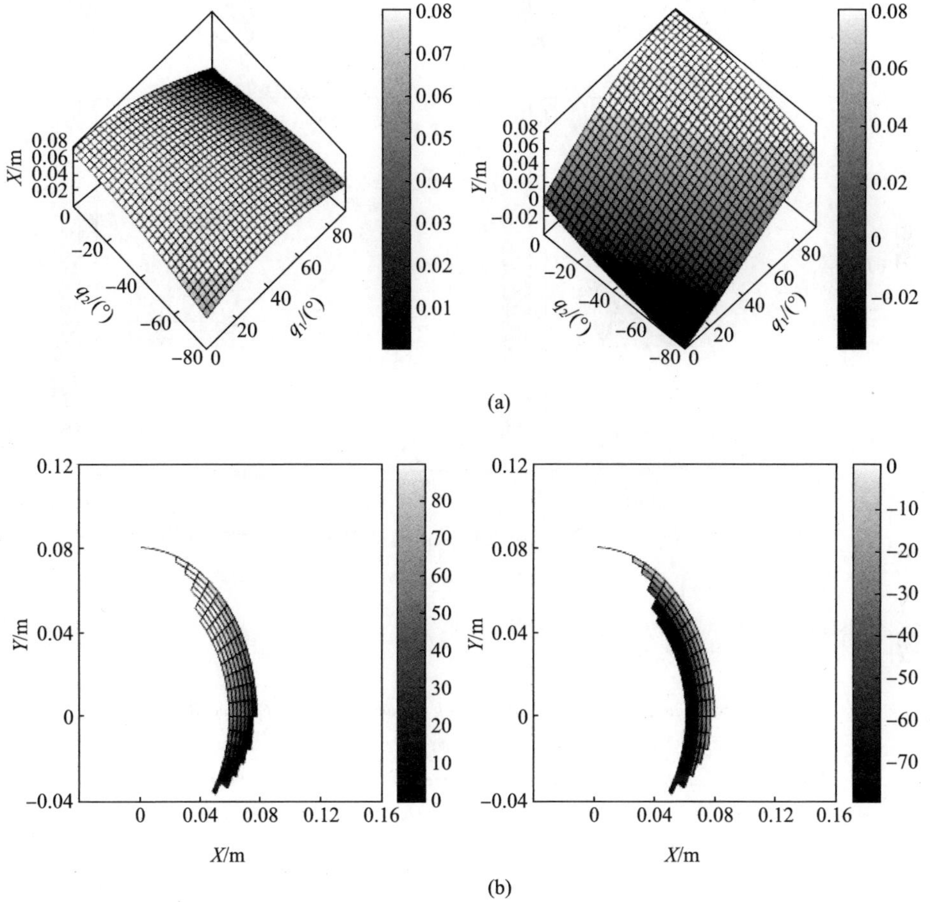

图 2.23　拇指的正向动力学和反向动力学

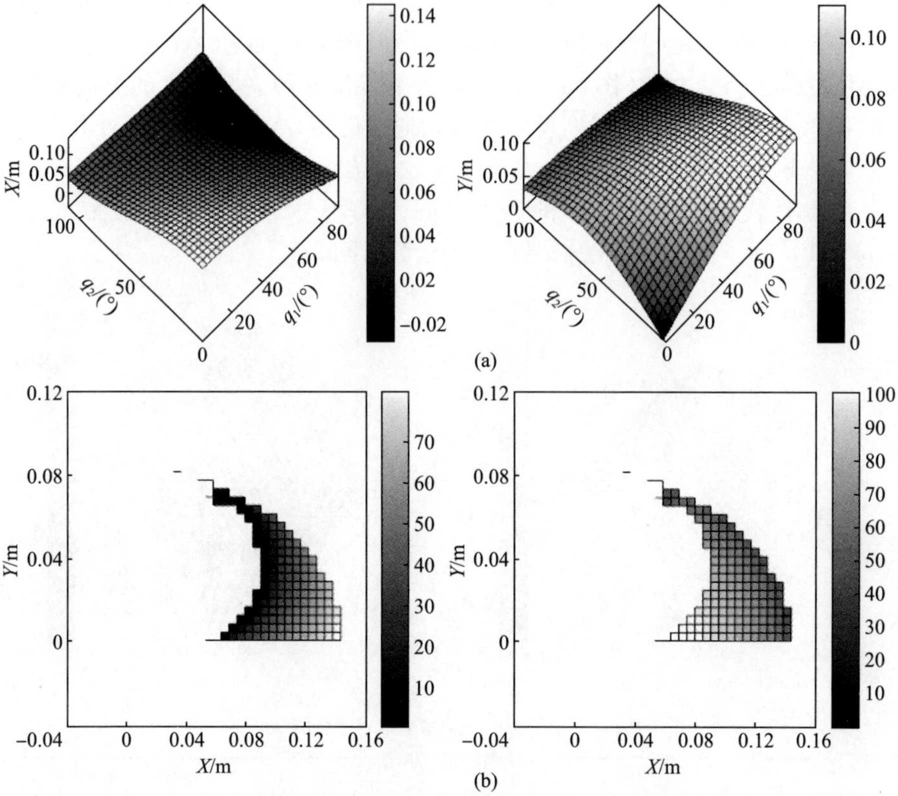

图 2.24　使用 ANFIS 法的食指正向动力学和反向动力学

参考文献

[1] R. R. Seeley, T. D. Stephens, and P. Tate. *Anatomy and Physiology, Eighth Edition*. The McGraw – Hill, New York, USA, 2007.

[2] C. – H. Chen, D. S. Naidu, and M. P. Schoen. Adaptive control for a five – fingered prosthetic hand with unknown mass and inertia. *World Scientific and Engineering Academy and Society (W SEAS) Journal on Systems*, 10(5): 148 – 161, May 2011.

[3] C. – H. Chen and D. S. Naidu. Hybrid control strategies for a five – finger robotic hand. *Biomedical Signal Processing and Control*, 8(4): 382 – 390, July 2013.

[4] C. – H. Chen. *Hybrid Control Strategies for Smart Prosthetic Hand*. PhD Dis – sertation, Idaho State University, Pocatello, Idaho, USA, May 2009.

[5] C. – H. Chen and D. S. Naidu. "Optimal control strategy for two – fingered smart prosthetic hand," in *Proceedings of the International Association of Science and Technology for Development (IASTED) International Conference on Robotics and Applications (RA 2010)*, pp. 190 – 196, Cambridge, Massachusetts, USA, November

1 – 3, 2010.

[6] R. J. Schilling. *Fundamentals of Robotics: Analysis and Control.* Prentice Hall, Englewood Cliffs, New Jersey, USA ; 1990.

[7] R. Kelly, V. Santibanez, and A. Loria. *Control of Robot Manipulators in Joint Space.* Springer, New York, USA, 2005.

[8] M. W. Spong, S. Hutchinson, and M. Vidyasagar. *Robot Dynamics and Control.* John Wiley & Sons, New York, USA, 2006.

[9] R. N. Jazar. *Theory of Applied Robotics. Kinematics, Dynamics, and Control.* Springer, New York, USA, 2007.

[10] B. Siciliano, L. Sciavicco, L. Villani, and G. Oriolo. *Robotics: Modelling, Planning and Control.* Springer – Verlag, London, UK, 2009.

[11] C. – H. Chen, K. W. Bosworth, M. P. Schoen, S. E. Bearden, D. S. Naidu, and A. Perez – Gracia. "A study of particle swarm optimization on leukocyte adhesion molecules and control strategies for smart prosthetic hand," in *2008 IEEE Swarm Intelligence Symposium* (*IEEE SIS08*), St. Louis, Missouri, USA, September 21 – 23, 2008.

[12] C. – H. Chen, D. S. Naidu, A. Perez – Gracia, and M. P. Schoen. "Fusion of hard and soft control techniques for prosthetic hand," in *Proceedings of the Inter – national Association of Science and Technology for Development* (*IASTED*) *International Conference on Intelligent Systems and Control* (*ISC 2008*), pp. 120 – 125, Orlando, Florida, USA, November 16 – 18, 2008.

[13] C. – H. Chen, D. S. Naidu, A. Perez – Gracia, and M. P. Schoen. "A hybrid adap – tive control strategy for a smart prosthetic hand," in *The 31st Annual Inter – national Conference of the IEEE Engineering Medicine and Biology Society* (*EMBS*), pp. 5056 – 5059, Minneapolis, Minnesota, USA, September 2 – 6, 2009.

[14] D. G. Kamper, E. G. Cruz, and M. P. Siegel. Stereotypical fingertip trajectories during grasp. *Journal of Neurophysiology*, 90: 3702 – 3710, 2003.

[15] X. Luo, T. Kline, H. C. Fisher, K. A. Stubblefield, R. V. Kenyon, and D. G. Kamper. "Integration of augmented reality and assistive devices for post – stroke hand opening rehabilitation," in *The International Conference of IEEE Engi – neering in Medicine and Biology Society* (*EMBS*), Shanghai, P. R. China, 2005.

[16] L. Zollo, S. Roccella, E. Guglielmelli, M. C. Carrozza, and P. Dario. Biomecha – tronic design and control of an anthropomorphic artificial hand for pros – thetic and robotic applications. *IEEE/ASME Transactions on Mechatronics*, 12(4): 418 – 429, August 2007.

[17] X. Wen, D. Sheng, and J. Huang. *A Hybrid Particle Swarm Optimization for Manipulator Inverse Kinematics Control*, volume 5226 of *Lecture Notes in Computer Science.* Springer – Verlag, Berlin, Germany, 2008.

[18] C. – H. Chen, D. S. Naidu, A. Perez – q acia, and M. P. Schoen. "A hybrid control strategy for five – fingered smart prosthetic hand," in *Joint 48th IEEE Conference on Decision and Control* (*CDC*) *and 28th Chinese Control Conference* (*CCC*), pp. 5102 – 5107, Shanghai, P. R. China, December 16 – 18, 2009.

[19] C. – H. Chen, D. S. Naidu, A. Perez – Gracia, and M. P. Schoen. "A hybrid optimal control strategy for a smart prosthetic hand," in *Proceedings of the ASME 2009 Dynamic Systems and Control Conference* (*DSCC*), Hollywood, Califor – nia, USA, October 12 – 14, 2009 (No. DSCC2009 – 2507).

[20] J. – S. R. Jang, C. – T. Sun, and E. Mizutani. *Neuro – Fuzzy and Soft Computing: A Computational Approach to Learning and Machine Intelligence.* Prentice Hall PTR, Upper Saddle River, New Jersey, USA, 1997.

[21] P. K. Lavangie and C. C. Norkin. *Joint Structure and Function: A Comprehen – sive Analysis, Third Edi-*

tion. F. A. Davis Company, Philadelphia, Pennsylvania, USA, 2001.

[22] S. Arimoto. *Control Theory of Multi - fingered Hands: A Modeling and Analytical - Mechanics Approach for Dexterity and Intelligence.* Springer - Verlag, London, UK, 2008.

[23] F. L. Lewis, D. M. Dawson, and C. T. Abdallah. *Robot Manipulators Control: Second Edition, Revised and Expanded.* Marcel Dekker, Inc., New York, USA, 2004.

[24] R. H. Bartels, J. C. Beatty, and B. A. Barsky. *Bezier Curves: Chapter 10in An Introduction to Splines for Use in Computer Graphics and Geometric Modelling.* Morgan Kaufmann, San Francisco, California, USA, 1998.

第3章 动态模型

电机常用来驱动小型机械手的关节,如机械手。传动装置是能够改变电机与驱动关节之间角速度的机械齿轮电缆。3.1 节通过推导直流电动机和机械齿轮之间执行器的数学模型来描述量化关系。在 3.2 节中,采用拉格朗日方法推导一个 n 个旋转关节串行连接的机械手的运动动力学方程。

3.1 制动器

3.1.1 直流电动机

对于较小的负载,如双关节拇指和三关节食指,通常使用直流电动机作为驱动器,如图 3.1 所示[1]。

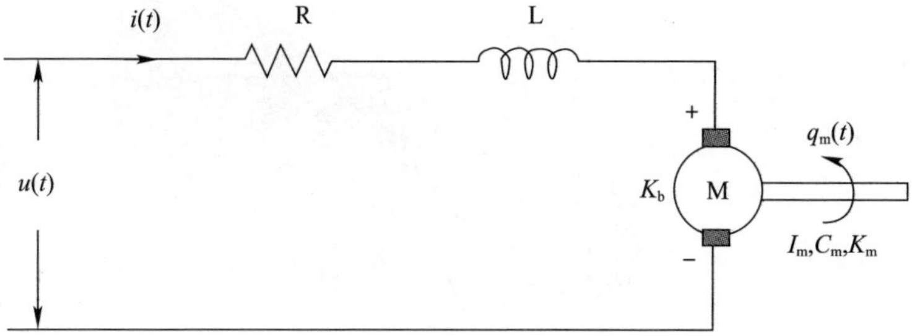

图 3.1 直流电动机

很容易为直流电动机电路写出如下关系:

$$L\frac{\mathrm{d}i}{\mathrm{d}t}+Ri(t)+K_{\mathrm{b}}\dot{q}(t)=u(t) \tag{3.1.1}$$

式中:R、L、K_{b} 分别是直流电动机电枢电阻、电感、反电动势常数。

电动机产生的力矩/力如下:

$$I_{\mathrm{m}}\ddot{q}(t)+C_{\mathrm{m}}\dot{q}(t)+K_{\mathrm{m}}q(t)=K_{\mathrm{b}}i(t) \tag{3.1.2}$$

式中:I_{m}、C_{m}、K_{m} 分别是电动机的惯性矩、阻尼系数、弹簧常数。对于双关节系统,我们需要有两个执行器。一般来说,在 n 关节系统中需要 n 个制动器,在这

种情况下,用如下矩阵形式改写式(3.1.1)和式(3.1.2):

$$\boldsymbol{L}\frac{\mathrm{d}\boldsymbol{i}}{\mathrm{d}t}+\boldsymbol{R}\boldsymbol{i}(t)+\boldsymbol{K}_{\mathrm{b}}\dot{\boldsymbol{q}}(t)=\boldsymbol{u}(t) \tag{3.1.3}$$

$$\boldsymbol{I}_{\mathrm{m}}\ddot{\boldsymbol{q}}_{\mathrm{m}}(t)+\boldsymbol{C}_{\mathrm{m}}\dot{\boldsymbol{q}}_{\mathrm{m}}(t)+\boldsymbol{K}_{\mathrm{m}}\boldsymbol{q}(t)=\boldsymbol{K}_{\mathrm{b}}\boldsymbol{i}(t) \tag{3.1.4}$$

式中:$\boldsymbol{q}(t)=\boldsymbol{q}_{\mathrm{m}}(t)$是电动机位移坐标矢量;$\boldsymbol{I}_{\mathrm{m}}\in\Re^{n\times n}$是一个包含电动机齿轮的对角正定惯性矩矩阵;$\boldsymbol{u}(t)\in\Re^{n}$ 和 $\boldsymbol{i}(t)\in\Re^{n}$ 分别代表电枢电压或控制输入和电流;$\boldsymbol{L}$、$\boldsymbol{R}$、$\boldsymbol{K}_{\mathrm{b}}$ 都是对角阵,分别代表电枢电感、电枢电阻和电动机反电动势常数。

3.1.2 机械齿轮传动

图3.2是机械齿轮传动示意图。图中:I_{m} 和 I 分别为电动机和负载关节对旋转轴的惯性矩;F_{m} 和 F 分别为电动机和负载关节的黏性摩擦因数;$\boldsymbol{\tau}_{\mathrm{m}}$ 和 $\boldsymbol{\tau}$ 分别为电动机的驱动力矩和负载关节的耦合力矩;q_{m} 和 q 分别为电动机和负载关节轴的角位置。若电动机齿轮和负载关节的齿轮没有滑动,则有

$$r_{\mathrm{m}}q_{\mathrm{m}}=rq \tag{3.1.5}$$

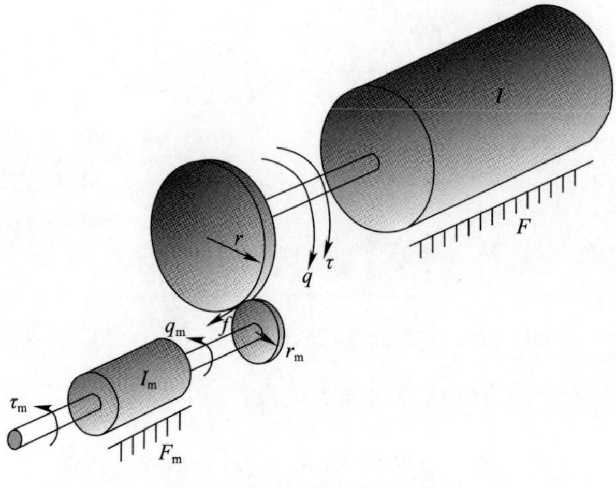

图3.2 机械齿轮传动示意图

式中:r_{m} 和 r 分别为电动机和负载关节的齿轮半径。齿轮减速比 k_r 定义如下:

$$k_r=\frac{r}{r_{\mathrm{m}}}=\frac{q_{\mathrm{m}}}{q}=\frac{\dot{q}_{\mathrm{m}}}{\dot{q}}=\frac{\ddot{q}_{\mathrm{m}}}{\ddot{q}} \tag{3.1.6}$$

式中：$\dot{q}_{\mathrm{m}}$和 $\dot{q}$ 分别为电动机和负载关节轴角速度；$\ddot{q}_{\mathrm{m}}$ 和 $\ddot{q}$ 分别为电动机和负载关节轴角加速度。如图 3.2 所示，在电动机齿轮和负载关节齿轮之间的接触交换力 f，对电动机轴会形成一个驱动转矩 fr_{m}，对负载关节轴会形成一个反作用力矩 fr，因此，电机轴和负载关节轴的力矩平衡可以表示为

$$\tau_{\mathrm{m}} = I_{\mathrm{m}}\ddot{q}_{\mathrm{m}} + F_{\mathrm{m}}\dot{q}_{\mathrm{m}} + fr_{\mathrm{m}} \tag{3.1.7}$$

$$f_r = I\ddot{q} + F\dot{q} + \tau \tag{3.1.8}$$

将式(3.1.6)代入式(3.1.8)得到如下关系：

$$\begin{aligned} f &= \frac{I\ddot{q} + F\dot{q} + \tau}{r} \\ &= \left(\frac{I}{r}\right)\left(\frac{\ddot{q}_{\mathrm{m}}}{k_r}\right) + \left(\frac{F}{r}\right)\left(\frac{\dot{q}_{\mathrm{m}}}{k_r}\right) + \frac{\tau}{r} \end{aligned} \tag{3.1.9}$$

将式(3.1.6)和式(3.1.9)代入式(3.1.7)得出如下关系：

$$\begin{aligned} \tau_{\mathrm{m}} &= I_{\mathrm{m}}\ddot{q}_{\mathrm{m}} + F_{\mathrm{m}}\dot{q}_{\mathrm{m}} + \left[\left(\frac{I}{r}\right)\left(\frac{\ddot{q}_{\mathrm{m}}}{k_r}\right) + \left(\frac{F}{r}\right)\left(\frac{\dot{q}_{\mathrm{m}}}{k_r}\right) + \frac{\tau}{r}\right] r_{\mathrm{m}} \\ &= I_{\mathrm{m}}\ddot{q}_{\mathrm{m}} + F_{\mathrm{m}}\dot{q}_{\mathrm{m}} + \left(\frac{I}{k_r^2}\right)\ddot{q}_{\mathrm{m}} + \left(\frac{F}{k_r^2}\right)\dot{q}_{\mathrm{m}} + \frac{\tau}{k_r} \\ &= \left(I_{\mathrm{m}} + \frac{I}{k_r^2}\right)\ddot{q}_{\mathrm{m}} + \left(F_{\mathrm{m}} + \frac{I}{k_r^2}\right)\dot{q}_{\mathrm{m}} + \frac{\tau}{k_r} \end{aligned} \tag{3.1.10}$$

式(3.1.10)表示了电机轴的驱动转矩 τ_{m} 和负载关节轴的反作用力矩 τ 之间的关系。

3.2 动 力 学

为了设计控制系统，需要建立一个数学模型来描述机械手的动态行为。本节用拉格朗日方法，利用动能和势能来推导机械手的动力学方程。机械手运动的拉格朗日方程如下[2-4]：

$$\begin{aligned} \frac{\mathrm{d}}{\mathrm{d}t}\left(\frac{\partial \mathcal{L}}{\partial \dot{\boldsymbol{q}}}\right) - \frac{\partial \mathcal{L}}{\partial \boldsymbol{q}} &= \boldsymbol{Q} \\ &= \boldsymbol{\tau} - \boldsymbol{F}_{\mathrm{v}}\dot{\boldsymbol{q}} - \boldsymbol{F}_{\mathrm{s}}\mathbf{sign}(\dot{\boldsymbol{q}}) - \boldsymbol{J}'\boldsymbol{F}_{\mathrm{ext}} \end{aligned} \tag{3.2.1}$$

式中：$\mathcal{L}$ 为拉格朗日算符；$\dot{\boldsymbol{q}}$ 和 $\boldsymbol{q}$ 分别为关节的角速度和角度矢量；$\boldsymbol{Q}$ 为合力矢

量；$\boldsymbol{\tau}$ 为给定的关节转矩矢量；$\boldsymbol{F}_{\mathrm{v}}$ 为黏性摩擦系数的对角正定矩阵；作为一个静摩擦力矩的简化模型，可以考虑库仑摩擦力矩 $\boldsymbol{F}_{\mathrm{s}}\mathbf{sign}(\dot{\boldsymbol{q}})$，其中 $\boldsymbol{F}_{\mathrm{s}}$ 为一个对角正定矩阵，$\mathbf{sign}(\dot{\boldsymbol{q}})$ 为一个矢量，其数值由单个关节的速度 sign 函数给出；$\boldsymbol{J}$ 为雅可比行列式；$\boldsymbol{F}_{\mathrm{ext}}$ 为环境中末端执行器施加的外部力矢量。拉格朗日算符 $\mathcal{L}$ 表示为

$$\mathcal{L} = T - V \tag{3.2.2}$$

式中：T 和 V 分别为动能和势能[5]。

3.3 双关节拇指

拇指的拉格朗日算符 $\mathcal{L}^{\mathrm{t}}$ 可以描述为

$$\mathcal{L}^{\mathrm{t}} = T^{\mathrm{t}} - V^{\mathrm{t}} \tag{3.3.1}$$

T^{t}和 V^{t}可以写为

$$\begin{aligned} T^{\mathrm{t}} &= \sum_{k=1}^{n=2} T_k^{\mathrm{t}} = \sum_{k=1}^{n=2} (T_k^{\mathrm{t,lin}} + T_k^{\mathrm{t,rot}}) \\ &= \sum_{k=1}^{n=2} \left(\frac{1}{2} m_k^{\mathrm{t}} \boldsymbol{v}_{\mathrm{c}k}^{\mathrm{tT}} \boldsymbol{v}_{\mathrm{c}k}^{\mathrm{t}} + \frac{1}{2} \boldsymbol{\omega}_k^{\mathrm{tT}} \boldsymbol{I}_k^{\mathrm{t}} \boldsymbol{\omega}_k^{\mathrm{t}} \right) \\ &= \sum_{k=1}^{n=2} \left(\frac{1}{2} m_k^{\mathrm{t}} \frac{\mathrm{d}}{\mathrm{d}t} \boldsymbol{p}_{\mathrm{c}k}^{\mathrm{tT}} \frac{\mathrm{d}}{\mathrm{d}t} \boldsymbol{p}_{\mathrm{c}k}^{\mathrm{t}} + \frac{1}{2} \boldsymbol{\omega}_k^{\mathrm{tT}} \boldsymbol{I}_k^{\mathrm{t}} \boldsymbol{\omega}_k^{\mathrm{t}} \right) \end{aligned} \tag{3.3.2}$$

$$V^{\mathrm{t}} = \sum_{k=1}^{n=2} V_k^{\mathrm{t}} \tag{3.3.3}$$

式中：n 为拇指的关节数；$\boldsymbol{T}_k^{\mathrm{t,lin}}$ 和 $\boldsymbol{T}_k^{\mathrm{t,rot}}$ 分别为动能的线性和旋转部分；$\boldsymbol{m}_k^{\mathrm{t}}$ 为关节 k 的质量；$\boldsymbol{v}_{\mathrm{c}k}^{\mathrm{t}}$为关节 k 的质心速度矢量；$\boldsymbol{p}_{\mathrm{c}k}^{\mathrm{t}}$为关节 k 的质心位置矢量；$\boldsymbol{\omega}_k^{\mathrm{t}}$ 为关节 k 的角速度矢量；$\boldsymbol{I}_k^{\mathrm{t}}$ 为关节 k 的惯性动差矩阵。$\boldsymbol{p}_{\mathrm{c}k}^{\mathrm{t}}$、$\boldsymbol{v}_{\mathrm{c}k}^{\mathrm{t}}$、$\boldsymbol{\omega}_k^{\mathrm{t}}$、$\boldsymbol{I}_k^{\mathrm{t}}$ 和 V_k^{t} 分别定义如下：

$$\begin{cases} \boldsymbol{p}_{\mathrm{c}1}^{\mathrm{t}} = \begin{bmatrix} l_1^{\mathrm{t}} \cos(q_1^{\mathrm{t}}) \\ l_1^{\mathrm{t}} \sin(q_1^{\mathrm{t}}) \\ 0 \end{bmatrix} \\ \boldsymbol{p}_{\mathrm{c}2}^{\mathrm{t}} = \begin{bmatrix} L_1^{\mathrm{t}} \cos(q_1^{\mathrm{t}}) + l_2^{\mathrm{t}} \cos(q_1^{\mathrm{t}} + q_2^{\mathrm{t}}) \\ L_1^{\mathrm{t}} \sin(q_1^{\mathrm{t}}) + l_2^{\mathrm{t}} \sin(q_1^{\mathrm{t}} + q_2^{\mathrm{t}}) \\ 0 \end{bmatrix} \end{cases} \tag{3.3.4}$$

$$\begin{cases} \boldsymbol{v}_{c1}^{t} = \dfrac{d}{dt}\boldsymbol{p}_{c1}^{t} = \dfrac{d}{dt}\begin{bmatrix} l_1^t\cos(q_1^t) \\ l_1^t\sin(q_1^t) \\ 0 \end{bmatrix} \\ \boldsymbol{v}_{c2}^{t} = \dfrac{d}{dt}\boldsymbol{p}_{c2}^{t} = \dfrac{d}{dt}\begin{bmatrix} L_1^t\cos(q_1^t) + l_2^t\cos(q_1^t + q_2^t) \\ L_1^t\sin(q_1^t) + l_2^t\sin(q_1^t + q_2^t) \\ 0 \end{bmatrix} \end{cases} \tag{3.3.5}$$

$$\begin{cases} \boldsymbol{\omega}_1^t = \dfrac{d}{dt}\begin{bmatrix} 0 \\ 0 \\ q_1^t \end{bmatrix} = \begin{bmatrix} 0 \\ 0 \\ \dot{q}_1^t \end{bmatrix} \\ \boldsymbol{\omega}_2^t = \dfrac{d}{dt}\begin{bmatrix} 0 \\ 0 \\ q_1^t + q_2^t \end{bmatrix} = \begin{bmatrix} 0 \\ 0 \\ \dot{q}_1^t + \dot{q}_2^t \end{bmatrix} \end{cases} \tag{3.3.6}$$

$$\begin{cases} \boldsymbol{I}_1^t = \begin{bmatrix} I_{xx1}^t & -I_{xy1}^t & -I_{xz1}^t \\ -I_{yx1}^t & I_{yy1}^t & -I_{yz1}^t \\ -I_{zx1}^t & -I_{zy1}^t & I_{zz1}^t \end{bmatrix} \\ \boldsymbol{I}_2^t = \begin{bmatrix} I_{xx2}^t & -I_{xy2}^t & -I_{xz2}^t \\ -I_{yx2}^t & I_{yy2}^t & -I_{yz2}^t \\ -I_{zx2}^t & -I_{zy2}^t & I_{zz2}^t \end{bmatrix} \end{cases} \tag{3.3.7}$$

$$\begin{cases} V_1^t = m_1^t g l_1^t \sin(q_1^t) \\ V_2^t = m_2^t g L_1^t \sin(q_1^t) + m_2^t g l_2^t \sin(q_1^t + q_2^t) \end{cases} \tag{3.3.8}$$

式中:$\boldsymbol{l}_k^t$ 表示前一关节的末端和关节质心的距离;L_k^t 为关节 k 的长度;q_k^t 为关节 k 的角度,它是一个时间函数;g 为重力加速度。对角元素 $I_{mnk}^t(k=1,2,m=n)$称为惯性极矩。

$$\begin{cases} I^{\mathrm{t}}_{xxk} = I^{\mathrm{t}}_{xk} = \int_{V_k} (y^2 + z^2)\,\mathrm{d}m \\ I^{\mathrm{t}}_{yyk} = I^{\mathrm{t}}_{yk} = \int_{V_k} (z^2 + x^2)\,\mathrm{d}m \\ I^{\mathrm{t}}_{zzk} = I^{\mathrm{t}}_{zk} = \int_{V_k} (x^2 + y^2)\,\mathrm{d}m \end{cases} \tag{3.3.9}$$

非对角元素 $\boldsymbol{I}^{\mathrm{t}}_{mnk}(k=1,2)$，$m \neq n$ 称为惯性积。

$$\begin{cases} I^{\mathrm{t}}_{xyk} = I^{\mathrm{t}}_{yxk} = \int_{V_k} (xy)\,\mathrm{d}m \\ I^{\mathrm{t}}_{yzk} = I^{\mathrm{t}}_{zyk} = \int_{V_k} (yz)\,\mathrm{d}m \\ I^{\mathrm{t}}_{zxk} = I^{\mathrm{t}}_{xzk} = \int_{V_k} (zx)\,\mathrm{d}m \end{cases} \tag{3.3.10}$$

式中：V_k 为关节 k 的机体区域。

因此，利用拉格朗日方法获得的拇指动态方程如下[6]：

$$\boldsymbol{M}(\boldsymbol{q})\ddot{\boldsymbol{q}} + \boldsymbol{C}(\boldsymbol{q}, \dot{\boldsymbol{q}}) + \boldsymbol{G}(\boldsymbol{q}) = \boldsymbol{\tau} - \boldsymbol{F}_{\mathrm{v}}\dot{\boldsymbol{q}} - \boldsymbol{F}_{\mathrm{s}}\mathbf{sign}(\dot{\boldsymbol{q}}) - \boldsymbol{J}'\boldsymbol{F}_{\mathrm{ext}} \tag{3.3.11}$$

或

$$\begin{bmatrix} M^{\mathrm{t}}_{11} & M^{\mathrm{t}}_{12} \\ M^{\mathrm{t}}_{21} & M^{\mathrm{t}}_{22} \end{bmatrix} \begin{bmatrix} \ddot{q}^{\mathrm{t}}_1 \\ \ddot{q}^{\mathrm{t}}_2 \end{bmatrix} + \begin{bmatrix} C^{\mathrm{t}}_1 \\ C^{\mathrm{t}}_2 \end{bmatrix} + \begin{bmatrix} G^{\mathrm{t}}_1 \\ G^{\mathrm{t}}_2 \end{bmatrix} = \begin{bmatrix} \tau^{\mathrm{t}}_1 \\ \tau^{\mathrm{t}}_2 \end{bmatrix} - \begin{bmatrix} F^{\mathrm{t}}_{\mathrm{v}1} & 0 \\ 0 & F^{\mathrm{t}}_{\mathrm{v}2} \end{bmatrix} \begin{bmatrix} \dot{q}^{\mathrm{t}}_1 \\ \dot{q}^{\mathrm{t}}_2 \end{bmatrix} -$$

$$\begin{bmatrix} F^{\mathrm{t}}_{\mathrm{s}1} & 0 \\ 0 & F^{\mathrm{t}}_{\mathrm{s}2} \end{bmatrix} \begin{bmatrix} \mathrm{sign}(\dot{q}^{\mathrm{t}}_1) \\ \mathrm{sign}(\dot{q}^{\mathrm{t}}_2) \end{bmatrix} - \begin{bmatrix} J^{\mathrm{t}}_{11} & J^{\mathrm{t}}_{12} \\ J^{\mathrm{t}}_{21} & J^{\mathrm{t}}_{22} \end{bmatrix} \begin{bmatrix} F^{\mathrm{t}}_x \\ F^{\mathrm{t}}_y \end{bmatrix} \tag{3.3.12}$$

式中

$$\begin{cases} M^{\mathrm{t}}_{11} = 2m^{\mathrm{t}}_2 L^{\mathrm{t}}_1 l^{\mathrm{t}}_2 \cos(q^{\mathrm{t}}_2) + m^{\mathrm{t}}_1 l^{\mathrm{t}2}_1 + m^{\mathrm{t}}_2 L^{\mathrm{t}2}_1 + m^{\mathrm{t}}_2 l^{\mathrm{t}2}_2 + I^{\mathrm{t}}_{zz1} + I^{\mathrm{t}}_{zz2} \\ M^{\mathrm{t}}_{12} = m^{\mathrm{t}}_2 L^{\mathrm{t}}_1 l^{\mathrm{t}}_2 \cos(q^{\mathrm{t}}_2) + m^{\mathrm{t}}_2 l^{\mathrm{t}2}_2 + I^{\mathrm{t}}_{zz2} \\ M^{\mathrm{t}}_{21} = M^{\mathrm{t}}_{12} \\ M^{\mathrm{t}}_{22} = m^{\mathrm{t}}_2 l^{\mathrm{t}2}_2 + I^{\mathrm{t}}_{zz2} \end{cases} \tag{3.3.13}$$

$$\begin{cases} C_1^t = -2m_2^t L_1^t l_2^t \sin(q_2^t)\dot{q}_1^t\dot{q}_2^t - m_2^t L_1^t l_2^t \sin(q_2^t)\dot{q}_2^t\dot{q}_2^t \\ C_2^t = m_2^t L_1^t l_2^t \sin(q_2^t)\dot{q}_1^t\dot{q}_1^t - m_2^t L_1^t l_2^t \sin(q_2^t)\dot{q}_1^t\dot{q}_2^t \end{cases} \tag{3.3.14}$$

$$\begin{cases} G_1^t = g(m_1^t l_1^t \cos(q_1^t)) + m_2^t L_1^t \cos(q_1^t) + m_2^t l_2^t \cos(q_1^t + q_2^t) \\ G_2^t = g m_2^t l_2^t \cos(q_1^t + q_2^t) \end{cases} \tag{3.3.15}$$

τ_1^t 和 τ_2^t 是关节 1 和关节 2 处的给定转矩;$\boldsymbol{M}(\boldsymbol{q})$是惯性矩阵;$\boldsymbol{C}(\boldsymbol{q},\dot{\boldsymbol{q}})$是科里奥利/向心矢量;$\boldsymbol{G}(\boldsymbol{q})$是重力矢量;$\boldsymbol{F}_v$ 是黏性摩擦系数的对角正定矩阵;$\boldsymbol{F}_s$ 是一个对角正定矩阵;$\mathbf{sign}(\dot{\boldsymbol{q}})$是一个矢量,其数值由单个关节的速度 sign 函数给出;$\boldsymbol{J}$ 是雅可比行列式;$\boldsymbol{F}_{ext}$是各个方向的外部力矢量。sign 函数为

$$\mathrm{sign}(x) = \begin{cases} 1, & x>0 \\ -1, & x<0 \end{cases}$$

式(3.3.11)可写作

$$\boldsymbol{M}(\boldsymbol{q})\ddot{\boldsymbol{q}} + \boldsymbol{N}(\boldsymbol{q},\dot{\boldsymbol{q}}) = \boldsymbol{\tau} \tag{3.3.16}$$

式中:$\boldsymbol{N}(\boldsymbol{q},\dot{\boldsymbol{q}}) = \boldsymbol{C}(\boldsymbol{q},\dot{\boldsymbol{q}}) + \boldsymbol{G}(\boldsymbol{q}) + \boldsymbol{F}_v\dot{\boldsymbol{q}} + \boldsymbol{F}_s\mathbf{sign}(\dot{\boldsymbol{q}}) + \boldsymbol{J}'\boldsymbol{F}_{ext}$,表示 $\boldsymbol{q}$ 和 $\dot{\boldsymbol{q}}$ 中的非线性量。

3.4 三关节食指

相似地,食指的动态方程[6]按照式(3.3.11)可以得到(通过 MAPLE 软件),即

$$\begin{bmatrix} M_{11}^i & M_{12}^i & M_{13}^i \\ M_{21}^i & M_{22}^i & M_{23}^i \\ M_{31}^i & M_{32}^i & M_{33}^i \end{bmatrix}\begin{bmatrix} \ddot{q}_1^i \\ \ddot{q}_2^i \\ \ddot{q}_3^i \end{bmatrix} + \begin{bmatrix} C_1^i \\ C_2^i \\ C_3^i \end{bmatrix} + \begin{bmatrix} G_1^i \\ G_2^i \\ G_3^i \end{bmatrix} = \begin{bmatrix} \tau_1^i \\ \tau_2^i \\ \tau_3^i \end{bmatrix} - \begin{bmatrix} F_{v1}^i & 0 & 0 \\ 0 & \mathrm{F}_{v2}^i & 0 \\ 0 & 0 & F_{v3} \end{bmatrix}$$

$$\begin{bmatrix} \dot{q}_1^i \\ \dot{q}_2^i \\ \dot{q}_3^i \end{bmatrix} - \begin{bmatrix} F_{s1}^i & 0 & 0 \\ 0 & F_{s2}^i & 0 \\ 0 & 0 & F_{s3}^i \end{bmatrix}\begin{bmatrix} \mathrm{sign}(\dot{q}_1^i) \\ \mathrm{sign}(\dot{q}_2^i) \\ \mathrm{sign}(\dot{q}_3^i) \end{bmatrix} - \begin{bmatrix} J_{11}^i & J_{21}^i \\ J_{12}^i & J_{22}^i \\ 0.7J_{12}^i & 0.7J_{22}^i \end{bmatrix}\begin{bmatrix} F_x^i \\ F_y^i \end{bmatrix} \tag{3.4.1}$$

式中

$$\begin{cases} M_{11}^{i} = 2m_2^{i}L_1^{i}l_2^{i}\sin(q_1^{i})\sin(q_1^{i}+q_2^{i}) + 2m_2^{i}L_1^{i}l_2^{i}\cos(q_1^{i})\cos(q_1^{i}+q_2^{i}) + \\ \qquad 2m_3^{i}L_1^{i}L_2^{i}\sin(q_1^{i})\sin(q_1^{i}+q_2^{i}) + \\ \qquad 2m_3^{i}L_1^{i}L_2^{i}\cos(q_1^{i})\cos(q_1^{i}+q_2^{i}) + \\ \qquad 2m_3^{i}L_1^{i}l_3^{i}\sin(q_1^{i})\sin(q_1^{i}+q_2^{i}+q_3^{i}) + \\ \qquad 2m_3^{i}L_1^{i}l_3^{i}\cos(q_1^{i})\cos(q_1^{i}+q_2^{i}+q_3^{i}) + \\ \qquad 2m_3^{i}L_2^{i}l_3^{i}\sin(q_1^{i}+q_2^{i})\sin(q_1^{i}+q_2^{i}+q_3^{i}) + \\ \qquad 2m_3^{i}L_2^{i}l_3^{i}\cos(q_1^{i}+q_2^{i})\cos(q_1^{i}+q_2^{i}+q_3^{i}) + \\ \qquad m_1^{i}l_1^{i2} + m_2^{i}L_1^{i2} + m_2^{i}l_2^{i2} + m_3^{i}L_1^{i2} + m_3^{i}L_2^{i2} + m_3^{i}l_3^{i2} + \\ \qquad I_{zz1}^{i} + I_{zz2}^{i} + I_{zz3}^{i} \\ M_{12}^{i} = m_2^{i}L_1^{i}l_2^{i}\sin(q_1^{i})\sin(q_1^{i}+q_2^{i}) + m_2^{i}L_1^{i}l_2^{i}\cos(q_1^{i})\cos(q_1^{i}+q_2^{i}) + \\ \qquad 2m_3^{i}L_2^{i}l_3^{i}\sin(q_1^{i}+q_2^{i})\sin(q_1^{i}+q_2^{i}+q_3^{i}) + \\ \qquad 2m_3^{i}L_2^{i}l_3^{i}\cos(q_1^{i}+q_2^{i})\cos(q_1^{i}+q_2^{i}+q_3^{i}) + \\ \qquad m_3^{i}L_1^{i}L_2^{i}\sin(q_1^{i})\sin(q_1^{i}+q_2^{i}) + m_3^{i}L_1^{i}L_2^{i}\cos(q_1^{i})\cos(q_1^{i}+q_2^{i}) + \\ \qquad m_3^{i}L_1^{i}l_3^{i}\sin(q_1^{i})\sin(q_1^{i}+q_2^{i}+q_3^{i}) + \\ \qquad m_3^{i}L_1^{i}l_3^{i}\cos(q_1^{i})\cos(q_1^{i}+q_2^{i}+q_3^{i}) + \\ \qquad m_2^{i}l_2^{i2} + m_3^{i}L_2^{i2} + m_3^{i}l_3^{i2} + I_{zz2}^{i} + I_{zz3}^{i} \\ M_{13}^{i} = m_3^{i}L_1^{i}l_3^{i}\sin(q_1^{i})\sin(q_1^{i}+q_2^{i}+q_3^{i}) + \\ \qquad m_3^{i}L_1^{i}l_3^{i}\cos(q_1^{i})\cos(q_1^{i}+q_2^{i}+q_3^{i}) + \\ \qquad m_3^{i}L_2^{i}l_3^{i}\sin(q_1^{i}+q_2^{i})\sin(q_1^{i}+q_2^{i}+q_3^{i}) + \\ \qquad m_3^{i}L_2^{i}l_3^{i}\cos(q_1^{i}+q_2^{i})\cos(q_1^{i}+q_2^{i}+q_3^{i}) + \\ \qquad m_3^{i}l_3^{i2} + I_{zz3}^{i} \end{cases} \tag{3.4.2}$$

$$\begin{cases} M_{21}^{i} = M_{12}^{i} \\ M_{22}^{i} = 2m_3^{i}L_2^{i}l_3^{i}\sin(q_1^{i}+q_2^{i})\sin(q_1^{i}+q_2^{i}+q_3^{i}) + \\ \qquad 2m_3^{i}L_2^{i}l_3^{i}\cos(q_1^{i}+q_2^{i})\cos(q_1^{i}+q_2^{i}+q_3^{i}) + \\ \qquad m_2^{i}l_2^{i2} + m_3^{i}l_2^{i2} + m_3^{i}l_3^{i2} + I_{zz2}^{i} + I_{zz3}^{i} \\ M_{23}^{i} = m_3^{i}L_2^{i}l_3^{i}\sin(q_1^{i}+q_2^{i})\sin(q_1^{i}+q_2^{i}+q_3^{i}) + \\ \qquad m_3^{i}L_2^{i}l_3^{i}\cos(q_1^{i}+q_2^{i})\cos(q_1^{i}+q_2^{i}+q_3^{i}) + \\ \qquad m_3^{i}l_3^{i2} + I_{zz3}^{i} \end{cases} \tag{3.4.3}$$

$$\begin{cases} M_{31}^{i} = M_{13}^{i} \\ M_{32}^{i} = M_{23}^{i} \\ M_{33}^{i} = m_3^{i} l_3^{i2} + I_{zz3}^{i} \end{cases} \tag{3.4.4}$$

$$\begin{cases} G_1^{i} = g(m_1^{i} l_1^{i} \cos(q_1^{i}) + m_2^{i} L_1^{i} \cos(q_1^{i}) + m_3^{i} L_1^{i} \cos(q_1^{i}) + \\ \qquad m_1^{i} l_2^{i} \cos(q_1^{i} + q_2^{i}) + m_3^{i} L_2^{i} \cos(q_1^{i} + q_2^{i}) + \\ \qquad m_3^{i} l_3^{i} \cos(q_1^{i} + q_2^{i} + q_3^{i})) \\ G_2^{i} = g(m_2^{i} l_2^{i} \cos(q_1^{i} + q_2^{i}) + m_3^{i} L_2^{i} \cos(q_1^{i} + q_2^{i}) + \\ \qquad m_3^{i} l_3^{i} \cos(q_1^{i} + q_2^{i} + q_3^{i})) \\ G_3^{i} = g(m_3^{i} l_3^{i} \cos(q_1^{i} + q_2^{i} + q_3^{i})) \end{cases} \tag{3.4.5}$$

$$\begin{aligned} C_1^{i} = {} & (2m_2^{i} L_1^{i} l_2^{i} \sin(q_1^{i}) \cos(q_1^{i} + q_2^{i}) - 2m_2^{i} L_1^{i} l_2^{i} \cos(q_1^{i}) \sin(q_1^{i} + q_2^{i}) + \\ & 2m_3^{i} L_1^{i} L_2^{i} \sin(q_1^{i}) \cos(q_1^{i} + q_2^{i}) - 2m_3^{i} L_1^{i} l_2^{i} \cos(q_1^{i}) \sin(q_1^{i} + q_2^{i}) + \\ & 2m_3^{i} L_1^{i} l_3^{i} \sin(q_1^{i}) \cos(q_1^{i} + q_2^{i} + q_3^{i}) - 2m_3^{i} L_1^{i} l_3^{i} \cos(q_1^{i}) \sin(q_1^{i} + q_2^{i} + q_3^{i})) \times \\ & \left(\frac{\partial q_1^{i}}{\partial t}\right)\left(\frac{\partial q_2^{i}}{\partial t}\right) + (2m_3^{i} L_1^{i} l_3^{i} \sin(q_1^{i}) \cos(q_1^{i} + q_2^{i} + q_3^{i}) - \\ & 2m_3^{i} L_1^{i} l_3^{i} \cos(q_1^{i}) \sin(q_1^{i} + q_2^{i} + q_3^{i}) + 2m_3^{i} L_2^{i} l_3^{i} \sin(q_1^{i} + q_2^{i}) \cos(q_1^{i} + q_2^{i} + q_3^{i}) - \\ & 2m_3^{i} L_2^{i} l_3^{i} \cos(q_1^{i} + q_2^{i}) \sin(q_1^{i} + q_2^{i} + q_3^{i})) \times \left(\frac{\partial q_1^{i}}{\partial t}\right)\left(\frac{\partial q_3^{i}}{\partial t}\right) + \\ & (2m_3^{i} L_1^{i} l_3^{i} \sin(q_1^{i}) \cos(q_1^{i} + q_2^{i} + q_3^{i}) - 2m_3^{i} L_1^{i} l_3^{i} \cos(q_1^{i}) \sin(q_1^{i} + q_2^{i} + q_3^{i}) + \\ & 2m_3^{i} L_2^{i} l_3^{i} \sin(q_1^{i} + q_2^{i}) \cos(q_1^{i} + q_2^{i} + q_3^{i}) - \\ & 2m_3^{i} L_2^{i} l_3^{i} \cos(q_1^{i} + q_2^{i}) \sin(q_1^{i} + q_2^{i} + q_3^{i})) \times \left(\frac{\partial q_2^{i}}{\partial t}\right)\left(\frac{\partial q_3^{i}}{\partial t}\right) + \\ & (m_2^{i} L_1^{i} l_2^{i} \sin(q_1^{i}) \cos(q_1^{i} + q_2^{i}) - m_2^{i} L_1^{i} l_2^{i} \cos(q_1^{i}) \sin(q_1^{i} + q_2^{i}) + \\ & m_3^{i} L_1^{i} L_2^{i} \sin(q_1^{i}) \cos(q_1^{i} + q_2^{i}) - m_3^{i} L_1^{i} L_2^{i} \cos(q_1^{i}) \sin(q_1^{i} + q_2^{i}) + \\ & m_3^{i} L_1^{i} l_3^{i} \sin(q_1^{i}) \cos(q_1^{i} + q_2^{i} + q_3^{i}) - m_3^{i} L_1^{i} l_3^{i} \cos(q_1^{i}) \sin(q_1^{i} + q_2^{i} + q_3^{i})) \times \\ & \left(\frac{\partial q_2^{i}}{\partial t}\right)\left(\frac{\partial q_2^{i}}{\partial t}\right) + (m_3^{i} L_1^{i} l_3^{i} \sin(q_1^{i}) \cos(q_1^{i} + q_2^{i} + q_3^{i}) - \\ & m_3^{i} L_1^{i} l_3^{i} \cos(q_1^{i}) \sin(q_1^{i} + q_2^{i} + q_3^{i}) + \end{aligned}$$

$$m_3^{\mathrm{i}} L_2^{\mathrm{i}} l_3^{\mathrm{i}} \sin(q_1^{\mathrm{i}} + q_2^{\mathrm{i}}) \cos(q_1^{\mathrm{i}} + q_2^{\mathrm{i}} + q_3^{\mathrm{i}}) -$$

$$m_3^{\mathrm{i}} L_2^{\mathrm{i}} l_3^{\mathrm{i}} \cos(q_1^{\mathrm{i}} + q_2^{\mathrm{i}}) \sin(q_1^{\mathrm{i}} + q_2^{\mathrm{i}} + q_3^{\mathrm{i}}) \times \left(\frac{\partial q_3^{\mathrm{i}}}{\partial t}\right)\left(\frac{\partial q_3^{\mathrm{i}}}{\partial t}\right)$$

$$C_2^{\mathrm{i}} = (m_2^{\mathrm{i}} L_1^{\mathrm{i}} l_2^{\mathrm{i}} \sin(q_1^{\mathrm{i}}) \cos(q_1^{\mathrm{i}} + q_2^{\mathrm{i}}) - m_2^{\mathrm{i}} L_1^{\mathrm{i}} l_2^{\mathrm{i}} \cos(q_1^{\mathrm{i}}) \sin(q_1^{\mathrm{i}} + q_2^{\mathrm{i}}) +$$

$$m_3^{\mathrm{i}} L_1^{\mathrm{i}} L_2^{\mathrm{i}} \sin(q_1^{\mathrm{i}}) \cos(q_1^{\mathrm{i}} + q_2^{\mathrm{i}}) - m_3^{\mathrm{i}} L_1^{\mathrm{i}} L_2^{\mathrm{i}} \cos(q_1^{\mathrm{i}}) \sin(q_1^{\mathrm{i}} + q_2^{\mathrm{i}}) +$$

$$m_3^{\mathrm{i}} L_1^{\mathrm{i}} l_3^{\mathrm{i}} \sin(q_1^{\mathrm{i}}) \cos(q_1^{\mathrm{i}} + q_2^{\mathrm{i}} + q_3^{\mathrm{i}}) - m_3^{\mathrm{i}} L_1^{\mathrm{i}} l_3^{\mathrm{i}} \cos(q_1^{\mathrm{i}}) \sin(q_1^{\mathrm{i}} + q_2^{\mathrm{i}} + q_3^{\mathrm{i}})) \times$$

$$\left(\frac{\partial q_1^{\mathrm{i}}}{\partial t}\right)\left(\frac{\partial q_2^{\mathrm{i}}}{\partial t}\right) + (2 m_3^{\mathrm{i}} L_2^{\mathrm{i}} l_3^{\mathrm{i}} \sin(q_1^{\mathrm{i}} + q_2^{\mathrm{i}}) \cos(q_1^{\mathrm{i}} + q_2^{\mathrm{i}} + q_3^{\mathrm{i}}) -$$

$$2 m_3^{\mathrm{i}} L_2^{\mathrm{i}} l_3^{\mathrm{i}} \cos(q_1^{\mathrm{i}} + q_2^{\mathrm{i}}) \sin(q_1^{\mathrm{i}} + q_2^{\mathrm{i}} + q_3^{\mathrm{i}})) \times \left(\frac{\partial q_1^{\mathrm{i}}}{\partial t}\right)\left(\frac{\partial q_3^{\mathrm{i}}}{\partial t}\right) +$$

$$(2 m_3^{\mathrm{i}} L_2^{\mathrm{i}} l_3^{\mathrm{i}} \sin(q_1^{\mathrm{i}} + q_2^{\mathrm{i}}) \cos(q_1^{\mathrm{i}} + q_2^{\mathrm{i}} + q_3^{\mathrm{i}}) -$$

$$2 m_3^{\mathrm{i}} L_2^{\mathrm{i}} l_3^{\mathrm{i}} \cos(q_1^{\mathrm{i}} + q_2^{\mathrm{i}}) \sin(q_1^{\mathrm{i}} + q_2^{\mathrm{i}} + q_3^{\mathrm{i}})) \times \left(\frac{\partial q_2^{\mathrm{i}}}{\partial t}\right)\left(\frac{\partial q_3^{\mathrm{i}}}{\partial t}\right) +$$

$$(-m_2^{\mathrm{i}} L_1^{\mathrm{i}} l_2^{\mathrm{i}} \sin(q_1^{\mathrm{i}}) \cos(q_1^{\mathrm{i}} + q_2^{\mathrm{i}}) + m_2^{\mathrm{i}} L_1^{\mathrm{i}} l_2^{\mathrm{i}} \cos(q_1^{\mathrm{i}}) \sin(q_1^{\mathrm{i}} + q_2^{\mathrm{i}}) +$$

$$m_3^{\mathrm{i}} L_1^{\mathrm{i}} L_2^{\mathrm{i}} \sin(q_1^{\mathrm{i}}) \cos(q_1^{\mathrm{i}} + q_2^{\mathrm{i}}) + m_3^{\mathrm{i}} L_1^{\mathrm{i}} L_2^{\mathrm{i}} \cos(q_1^{\mathrm{i}}) \sin(q_1^{\mathrm{i}} + q_2^{\mathrm{i}}) -$$

$$m_3^{\mathrm{i}} L_1^{\mathrm{i}} l_3^{\mathrm{i}} \sin(q_1^{\mathrm{i}}) \cos(q_1^{\mathrm{i}} + q_2^{\mathrm{i}} + q_3^{\mathrm{i}}) + m_3^{\mathrm{i}} L_1^{\mathrm{i}} l_3^{\mathrm{i}} \cos(q_1^{\mathrm{i}}) \sin(q_1^{\mathrm{i}} + q_2^{\mathrm{i}} + q_3^{\mathrm{i}})) \times$$

$$\left(\frac{\partial q_1^{\mathrm{i}}}{\partial t}\right)\left(\frac{\partial q_1^{\mathrm{i}}}{\partial t}\right) + (m_3^{\mathrm{i}} L_2^{\mathrm{i}} l_3^{\mathrm{i}} \sin(q_1^{\mathrm{i}} + q_2^{\mathrm{i}}) \cos(q_1^{\mathrm{i}} + q_2^{\mathrm{i}} + q_3^{\mathrm{i}}) -$$

$$m_3^{\mathrm{i}} L_2^{\mathrm{i}} l_3^{\mathrm{i}} \cos(q_1^{\mathrm{i}} + q_2^{\mathrm{i}}) \sin(q_1^{\mathrm{i}} + q_2^{\mathrm{i}} + q_3^{\mathrm{i}})) \times \left(\frac{\partial q_3^{\mathrm{i}}}{\partial t}\right)\left(\frac{\partial q_3^{\mathrm{i}}}{\partial t}\right)$$

$$C_3^{\mathrm{i}} = (2 m_3^{\mathrm{i}} L_2^{\mathrm{i}} l_3^{\mathrm{i}} \cos(q_1^{\mathrm{i}} + q_2^{\mathrm{i}}) \sin(q_1^{\mathrm{i}} + q_2^{\mathrm{i}} + q_3^{\mathrm{i}}) -$$

$$2 m_3^{\mathrm{i}} L_2^{\mathrm{i}} l_3^{\mathrm{i}} \sin(q_1^{\mathrm{i}} + q_2^{\mathrm{i}}) \cos(q_1^{\mathrm{i}} + q_2^{\mathrm{i}} + q_3^{\mathrm{i}})) \times \left(\frac{\partial q_1^{\mathrm{i}}}{\partial t}\right)\left(\frac{\partial q_2^{\mathrm{i}}}{\partial t}\right) +$$

$$(m_3^{\mathrm{i}} L_1^{\mathrm{i}} l_3^{\mathrm{i}} \sin(q_1^{\mathrm{i}}) \cos(q_1^{\mathrm{i}} + q_2^{\mathrm{i}} + q_3^{\mathrm{i}}) - m_3^{\mathrm{i}} L_1^{\mathrm{i}} l_3^{\mathrm{i}} \cos(q_1^{\mathrm{i}}) \sin(q_1^{\mathrm{i}} + q_2^{\mathrm{i}} + q_3^{\mathrm{i}}) +$$

$$(m_3^{\mathrm{i}} L_2^{\mathrm{i}} l_3^{\mathrm{i}} \sin(q_1^{\mathrm{i}} + q_2^{\mathrm{i}}) \cos(q_1^{\mathrm{i}} + q_2^{\mathrm{i}} + q_3^{\mathrm{i}}) -$$

$$m_3^{\mathrm{i}} L_2^{\mathrm{i}} l_3^{\mathrm{i}} \cos(q_1^{\mathrm{i}} + q_2^{\mathrm{i}}) \sin(q_1^{\mathrm{i}} + q_2^{\mathrm{i}} + q_3^{\mathrm{i}})) \times \left(\frac{\partial q_1^{\mathrm{i}}}{\partial t}\right)\left(\frac{\partial q_3^{\mathrm{i}}}{\partial t}\right) +$$

$$(m_3^{\mathrm{i}} L_2^{\mathrm{i}} l_3^{\mathrm{i}} \sin(q_1^{\mathrm{i}} + q_2^{\mathrm{i}}) \cos(q_1^{\mathrm{i}} + q_2^{\mathrm{i}} + q_3^{\mathrm{i}}) -$$

$$m_3^{\mathrm{i}} L_2^{\mathrm{i}} l_3^{\mathrm{i}} \cos(q_1^{\mathrm{i}} + q_2^{\mathrm{i}}) \sin(q_1^{\mathrm{i}} + q_2^{\mathrm{i}} + q_3^{\mathrm{i}})) \times \left(\frac{\partial q_2^{\mathrm{i}}}{\partial t}\right)\left(\frac{\partial q_3^{\mathrm{i}}}{\partial t}\right) +$$

$$(m_3^{\mathrm{i}} L_1^{\mathrm{i}} l_3^{\mathrm{i}} \cos(q_1^{\mathrm{i}}) \sin(q_1^{\mathrm{i}} + q_2^{\mathrm{i}} + q_3^{\mathrm{i}}) - m_3^{\mathrm{i}} L_1^{\mathrm{i}} l_3^{\mathrm{i}} \sin(q_1^{\mathrm{i}}) \cos(q_1^{\mathrm{i}} + q_2^{\mathrm{i}} + q_3^{\mathrm{i}}) +$$

$$m_3^{\mathrm{i}} L_2^{\mathrm{i}} l_3^{\mathrm{i}} \cos(q_1^{\mathrm{i}} + q_2^{\mathrm{i}}) \sin(q_1^{\mathrm{i}} + q_2^{\mathrm{i}} + q_3^{\mathrm{i}}) -$$

$$m_3^{\mathrm{i}} L_2^{\mathrm{i}} l_3^{\mathrm{i}} \sin(q_1^{\mathrm{i}} + q_2^{\mathrm{i}}) \cos(q_1^{\mathrm{i}} + q_2^{\mathrm{i}} + q_3^{\mathrm{i}})) \times \left(\frac{\partial q_1^{\mathrm{i}}}{\partial t}\right)\left(\frac{\partial q_1^{\mathrm{i}}}{\partial t}\right) +$$

$$(m_3^{\mathrm{i}} L_2^{\mathrm{i}} l_3^{\mathrm{i}} \cos(q_1^{\mathrm{i}} + q_2^{\mathrm{i}}) \sin(q_1^{\mathrm{i}} + q_2^{\mathrm{i}} + q_3^{\mathrm{i}}) -$$

$$m_3^{\mathrm{i}} L_2^{\mathrm{i}} l_3^{\mathrm{i}} \sin(q_1^{\mathrm{i}} + q_2^{\mathrm{i}}) \cos(q_1^{\mathrm{i}} + q_2^{\mathrm{i}} + q_3^{\mathrm{i}})) \times \left(\frac{\partial q_2^{\mathrm{i}}}{\partial t}\right)\left(\frac{\partial q_2^{\mathrm{i}}}{\partial t}\right) \tag{3.4.6}$$

参考文献

[1] B. C. Kuo. *Automatic Control Systems, Seventh Edition.* Prentice Hall , Engle – wood Cliffs, New Jersey, USA,1995.

[2] R. Krlly, V. Santibanez, and A. Loria. *Control of Robot Manipulators in Joint space.* Springer, New York, USA, 2005.

[3] R. N. Jazar. *Theory of Applied Robotics. Kinematics, Dynamics, and Control.* Springer, New York, USA, 2007.

[4] B. Siciliano, L. Sciavicco, L. Villani, and G. Oriolo. *Robotics: Modelling, Planning and Control.* Springer – Verlag, London, UK, 2009.

[5] C. – H. Chen. *Hybrid Control Strategies for Smart Prosthetic Hand.* PhD Dissertation, Idaho State University, Pocatello, Idaho, USA, May 2009.

[6] C. – H. Chen, D. S. Naidu, and M. P. Schoen. Adaptive control for a five – fingered prosthetic hand with unknown mass and inertia. *World Scientific and Engineering Academy and Society (WSEAS) Journal on Systems*, 10(5): 148 – 161, May 2011.

第4章　软计算/控制策略

硬控制策略提供了固定的解决方案去控制动态系统。当需要处理没有被清晰定义或获知的任务时,基于知识的系统更适合去完成这类任务[1]。本章主要介绍一些软计算(SC)或计算智能(CI)[2]策略,其中包括4.1节模糊逻辑(FL),4.2节神经网络(NN),4.3节自适应神经模糊推理系统(ANFIS),4.4节禁忌算法(TS),4.5节遗传算法(GA),4.6节粒子群优化(PSO),4.7节自适应粒子群优化(APSO)以及4.8节凝聚混合优化(CHO)。所有的仿真结果及分析在4.9节给出。

4.1　模糊逻辑

人类是灵活的且可以适应不熟悉的情况,另外人类还可以用有效的方式得到有用的信息,同时丢弃其中没有价值的细节。对于人类,生成的信息不需要完整或者精准,这些信息可能是近似的、模糊的且定性的,在进行总结推理后,得到新的信息和知识[1]。一个典型的小于30的实数集合A可以写为

$$A = \{x \mid x < 30\} \tag{4.1.1}$$

式中:30是一个边界数值,如果x小于该数值,那么x属于集合A;否则,x不属于集合A。设定x和A分别代表一个人的年龄和年轻程度,如果Chen的年龄是29.99,这个年龄小于30,则属于集合A,那么人们会认为Chen是年轻的(A);然而,如果Mario的年龄为30.01,其大于30,则不属于集合A,但在实际生活中,人们不会认为Mario不再年轻了。换句话说,29.99和30.01之间的差别并不明显。与传统集合相比,模糊集合就是一个没有清晰边界或没有二值特征的集合。即"属于这个集合"与"不属于这个集合"的界定是平缓且有梯度的。图4.1和4.2是"Mario是年轻的"和"Chen是成绩好的"有关模糊集合的简单例子。

模糊集合是对不确定性事物进行数学建模而得到的数学对象,利用隶属函数度的概念可以对一个模糊集合进行数学定义。若X是一个由x表示的集合,那么一个属于X的模糊集合A可以由一对元素构成的集合来定义:

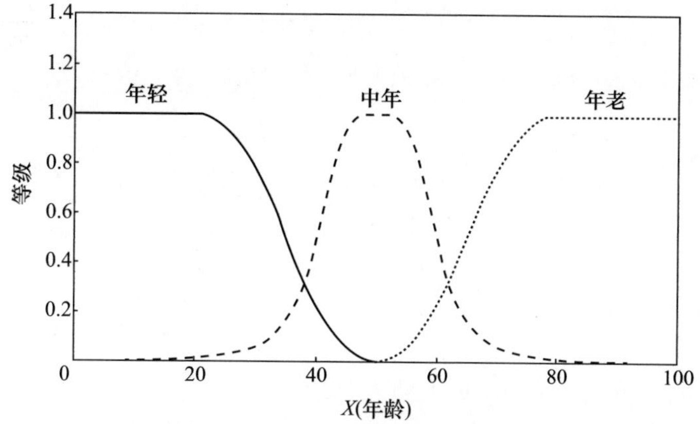

图 4.1 “Mario 是年轻的”隶属函数

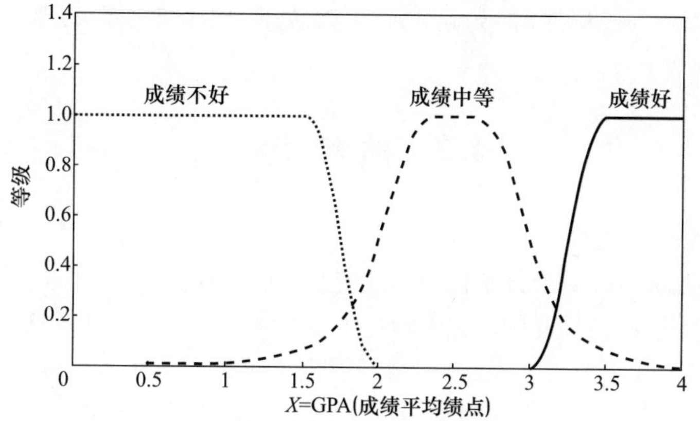

图 4.2 “Chen 是成绩好的”隶属函数

$$A = \{ (x, \mu_A(x)) \mid x \in X \cap \mu_A(x) \in [0,1] \} \tag{4.1.2}$$

式中：隶属函数（MF）$\mu_A(x)$ 表示元素 x 属于模糊集合 A 的可能性。

图 4.1 和图 4.2利用 Z 形隶属函数 $\mu_z(x)$、广义钟形隶属函数 $\mu_g(x)$ 和 S 形隶属函数 $\mu_s(x)$ 三种隶属函数分别表示年轻/成绩不好、年龄/成绩适中和年老/成绩好这三种情况。这三种隶属函数可以表示为

$$\mu_z(x) = \begin{cases} 1, & x \leqslant a_z \\ 1 - 2\left(\dfrac{x - a_z}{b_z - a_z}\right)^2, & a_z \leqslant x \leqslant \dfrac{a_z + b_z}{2} \\ 2\left(\dfrac{b_z - x}{b_z - a_z}\right)^2, & \dfrac{a_z + b_z}{2} \leqslant x \leqslant b_z \\ 0, & x \geqslant b_z \end{cases}$$

$$\mu_g(x) = \frac{1}{1 + \left|\frac{x - c_g}{a_g}\right|^{2b_g}}$$

$$\mu_s(x) = \begin{cases} 1, & x \leqslant a_s \\ 2\left(\frac{x - a_s}{b_s - a_s}\right)^2, & \frac{a_s + b_s}{2} \leqslant x \leqslant b_s \\ 1 - 2\left(\frac{b_s - x}{b_s - a_s}\right)^2, & a_s \leqslant x \leqslant \frac{a_s + b_z}{2} \\ 0, & x \geqslant b_s \end{cases}$$

式中:参数 a_z、b_z、a_g、b_g、c_g、a_s、b_s 是可选择的常量。

此外,人类可以根据“如果……,那么……”这种规则做很多事情。比如:如果温度很低,那么就升高温度;如果累了,那么就去睡觉。模糊规则“如果-那么”(也可以称为模糊规则/模糊影响)可以表述为“如果 x 是 A,那么 y 是 B”的形式。

4.2 神经网络

生物神经元是所有人工神经网络(ANN)的基础,后文简称为神经网络(NN)。生物神经元是由细胞体、树突、突触和轴突组成[3]。轴突是一个具有很多分支的管状物,神经元的轴突末梢终止于突触小体处,单个神经细胞的轴突与其他很多神经元形成了突触连接。第一个神经元的突触小体与另外一个神经元的受体位点相距一个微小的距离,第一个神经元的细胞体产生一种称为神经递质的化学物质,这些物质被运输到突触小泡中。神经递质一直被存储在突触小泡中,直到神经元被激活。突触小泡将神经递质释放,神经递质流过突触间隙,并作用在第二个神经元上。因此,第一个神经元的神经递质可以刺激或者抑制第二个神经元的活动,每个神经元都会接收来自成百上千个其他神经元的信息。

以上内容表明:树突可以通过突触连接从其他神经元接收信号(电动势),并传入细胞体,细胞体在接收到该信号后会进行信号整合以及信号再传递。一个神经元的细胞体增加自树突传递而来的信号,如果输入信号的强度达到一个阈值,那么这个神经元就会被激活,并且发送一个信号给它的轴突;如果输入信号的强度达不到所需的阈值水平,那么输入信号就会迅速衰减,这个神经元不会对轴突产生任何作用[3]。

受生物神经系统的启发,神经网络(NN)通常由一组并行且具有分布式处理信息能力的单元组成,这些单元称为节点或神经元,它们排序成层并且通过单相加权信号通道建立起合适的内部联系,这种内部联系称为连接或者突触权重[1,4-6]。神经网络中的所有节点或部分节点可以是自适应的,换句话说,节点的输出会修改与这些节点相关的参数。通过神经网络的学习规则会更新这些参数,从而使得“规定误差”最小,这个规定误差通常是指期望输出和实际输出的差值。图 4.3 呈现了一个带有输入层、隐藏层和输出层的简单前馈神经网络模型。

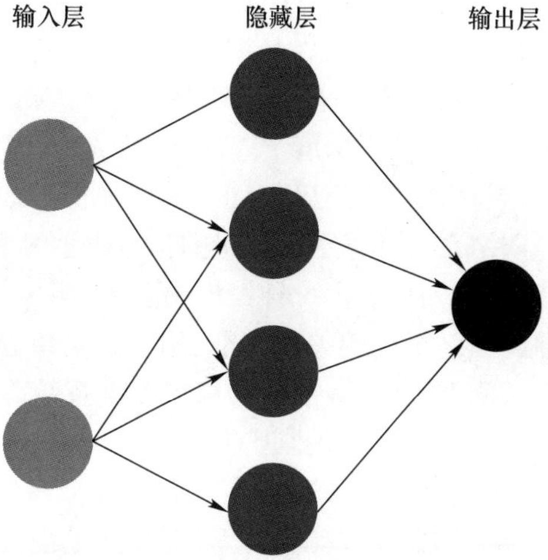

图 4.3　一个简单的前馈人工神经网络

4.3　自适应神经模糊推理系统

1965 年,L. A. Zadel 首先提出了一个以隶属函数为特征的模糊集合[7]。不同于传统集合,模糊集合的隶属度都被定义在 0 到 1 之间。利用模糊集合,模糊逻辑(FL)提出一种数学模型来描述不确定环境和近似知识推理。模糊逻辑控制器(FLC)是由模糊集合和模糊逻辑基础上发展而来的,模糊逻辑控制器利用模糊“如果 - 那么”规则(也称为模糊规则)控制系统。该规则中“如果”这部分作为先验知识,“那么”部分作为推论[1,8]。在模糊逻辑控制器的处理方法中,最常用的模型是 Mamdani 模型[9] 和 Sugeno(或 Takagi - Sugeno - Kang, TSK)模型[10]。这两个模型使用同种处理方式将传统输入模糊化,并且都应用了模糊运算符。Mamdani 模型和 Sugeno 模型之间的主要区别是,Sugeno 类型使用线性或者常数(例如一阶或零阶多项式函数)得到输出而不使用隶属函数。

神经网络是一个具有模型结构的算法，它可以学习、训练以及调整权重参数来满足给定的非线性数据。受生物神经系统的启发，神经网络通常由一系列并行的且具有分布式处理能力的单元（称作节点或者神经元）组成。这些节点通常有序地放置在层中（包括输入层、隐藏层、输出层），这些层通过单相加权信号通道（称作连接或者突触权重）[1,4,11]建立适当的内部联系。神经网络中的全部或者部分节点通过使用学习规则和算法来调整权重参数，比如最小方差（LSE，也称作 Widrow – Hoff 学习规则）[12]、梯度下降算法等。也就是说，节点的输出可以调整与这些节点相关的权重参数并且学习规则可以更新这些参数，以最小化规定的误差测量，即网络的实际输出与期望/目标输出之间的差异，或者改变规定的误差测量梯度的学习速率。神经网络利用自适应 LSE 算法在强非线性系统中拟合给定的数据，并且利用梯度下降法来控制收敛速度。

FL 系统利用模糊集合来建模不确定性和近似知识推理，但是 FL 体系架构缺乏学习规则。神经网络系统加强了数值集的自适应学习规则以测试非线性数据，但是神经网络特征缺乏知识表示。神经模糊系统（NFS）包括了智能系统，它结合了模糊逻辑和神经网络系统的主要特征，用来解决单独使用模糊逻辑或神经网络方法无法解决的问题[13]。最常用的神经模糊系统是自适应神经模糊推理系统（ANFIS）[14]，自适应神经模糊推理系统是一个嵌入自适应网络框架的模糊推理系统，它提供最佳优化算法来确定参数以适应给定的数据。基于人类的推理，在模糊“如果 – 那么”规则框架下，自适应神经模糊推理系统使用混合学习过程建立了输入 – 输出数据对的映射。

双关节拇指的逆运动学解析解可以通过数学方法推理得到。然而，对于更复杂的结构（如增加三维空间的自由度），这将成为一个难以解决的问题。由于三关节食指的正向运动学是公式化的[15]，笛卡尔坐标中的指尖工作空间是由所有关节的旋转角度的整个范围建立的。因此，三关节食指的逆动力学问题可以使用自适应神经模糊推理系统解决[4,14]，其中笛卡尔空间作为输入，关节空间作为输出。笛卡尔空间（输入）和关节空间（输出）作为训练数据集存储，然后由自适应神经模糊推理系统训练，自适应神经模糊推理系统包含前提参数（也称作先行参数），前文定义的隶属函数以及结果参数，这样可以确定每个输出方程的系数。在自适应神经模糊推理系统的混合学习过程中，在后向通道中使用反向传播梯度下降算法来调整隶属函数的前提参数，在前向通道中使用 LSE（也称作 Widrow – Hoff 学习规则）[12]方法来调整输出函数的结果参数。

J. – S. Jang 提出三种类型的自适应神经模糊推理系统[14]，我们使用类型 3 的自适应神经模糊推理系统，这个系统使用 Takagi – Sugeno 的“如果 – 那么”模糊规则，此规则下的输出是一个输入变量和常数的线性结合。简单地总结类型

3 的自适应神经模糊推理系统的结构，假设 x 和 y 是两个输入变量，$f(x,y)$ 是一个输出变量。如图 4.4(a)所示，对于一个一阶 Sugeno 模糊模型，两个 Takagi - Sugeno 的“如果 - 那么”模糊规则可以表示如下：

规则 1：如果 x 是 A_1 并且 y 是 B_1，那么 $f_1 = p_1x + q_1y + r_1$。

规则 2：如果 x 是 A_2 并且 y 是 B_2，那么 $f_2 = p_2x + q_2y + r_2$。

这里，A_i 和 $B_i(i=1,2)$ 是语言标签（例如小、中和大），p_i、q_i 和 $r_i(i=1,2)$ 是线性结果参数，图 4.4(b)描述了五层等效自适应神经模糊推理系统架构，这五层分别称作模糊化（第一层）、乘积（第二层）、标准化（第三层）、去模糊化（第四层）、聚合（第五层）[16-17]。三关节食指的逆运动学问题可以使用以笛卡尔空间作为输入、关节空间作为输出的自适应神经模糊推理系统求解。笛卡尔空间

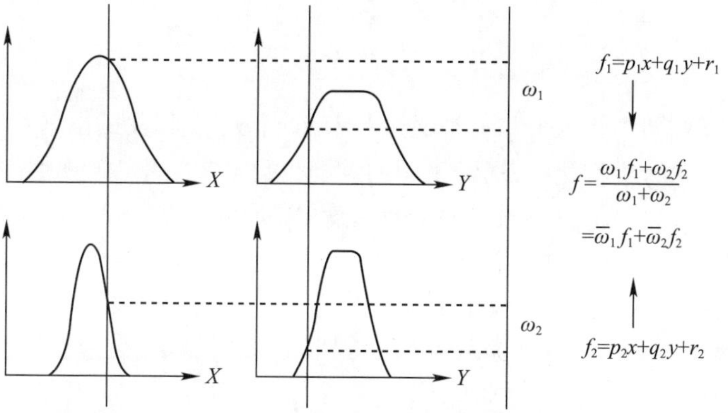

(a) 基于两个“如果-那么”模糊规则的双输入变量x和y，单输出变量 f 的一阶Sugeno模糊模型

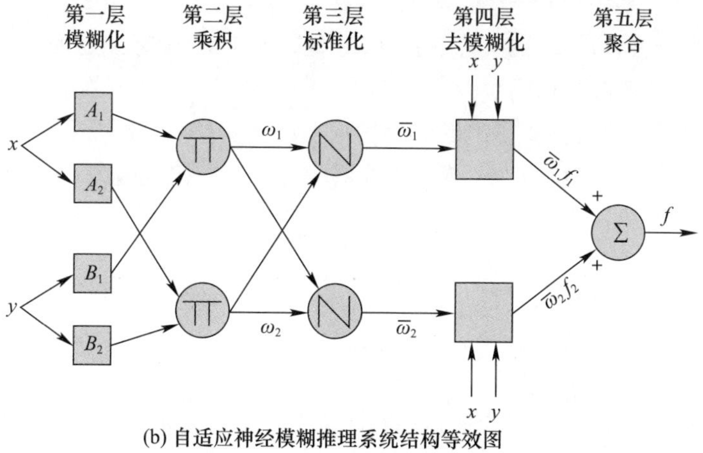

(b) 自适应神经模糊推理系统结构等效图

图 4.4　自适应神经模糊推理系统结构

(输入)和关节空间(输出)作为训练数据集存储,然后由自适应神经模糊推理系统训练。在自适应神经模糊推理系统混合学习过程中在向后传递过程调优隶属函数(第一层)的前提参数时,使用了反向传播梯度下降法;在向前传递过程中调整输出函数(第四层)的结果参数时,使用了最小方差(LSE,也是 Widrow - Hoff 学习规则)[4,14]。

第一层:模糊化层。

定义 $O_{i,j}$ 是第 i 层第 j 个节点的输出,方形节点($i=1,4$)是带有参数的自适应节点,圆形节点($i=2,3,5$)是没有参数的固定节点。在层 $i(i=1)$ 的节点 $j(j=1,2)$ 是带有节点函数的方形节点。

$$\begin{aligned} O_{1,j} &= \mu_{A_j}(x) \\ O_{1,j} &= \mu_{B_j}(y) \end{aligned} \tag{4.3.1}$$

节点 j 的两个输入 x 和 y 是通过与节点函数 $O_{1,j}$ 相关的语言标签的隶属函数 μ_{A_j} 和 μ_{B_j} 进行模糊化的,通常采用的隶属函数是三角型、梯型、高斯型、钟型等值在[0,1]区间的函数。例如给出钟形隶属函数:

$$\mu_{A_j}(x) = \frac{1}{1 + \left[\left(\frac{x - c_j}{a_j}\right)^2\right]^{b_j}} \tag{4.3.2}$$

这里参数集 $\{a_j, b_j, c_j\}$ 包括了前提参数。

第二层:乘积(范数运算符)层。

在这个层 $i(i=2)$ 的节点 $j(j=1,2)$ 是图 4.4(b)中标记为乘积运算符 $\prod$(或 T 范数 $\otimes$)且带有节点函数 $O_{2,j}$ 的圆形(固定)节点,这个层的所有节点将所有输入信号做乘法并将乘积得出的结果发送至下一层(第三层),第三层表示模糊先验规则的激励强度(“如果”部分),这层的输出充当权重函数 ω_j,可以表示为

$$O_{2,j} = \omega_j = \mu_{A_j}(x) \otimes \mu_{B_j}(y) \tag{4.3.3}$$

第三层:标准化层。

在 $i(i=3)$ 层的节点 $j(j=1,2)$ 是图 4.4(b)中标记为 N 且具有节点函数 $O_{3,j}$ 的圆形节点。在这层的第 j 个节点计算了第 j 个规则产生的激励强度占所有规则产生的激励强度的比例,这个层的输出将权重函数标准化,其中,权重函数由之前的乘积层传递而来。标准化后,权重函数(激励强度)$\bar{\omega}_j$ 可以写为

$$O_{3,j} = \bar{\omega}_j = \frac{\omega_j}{\sum_j \omega_j} \tag{4.3.4}$$

第四层:去模糊化(结果)层。

这个层$i(i=4)$的节点j是带有节点函数$O_{4,j}$的方形节点,该层第j个节点根据模糊结果规则("那么"部分)进行去模糊化。去模糊化的结果与标准化的激励强度进行乘积运算,公式为

$$O_{4,j} = \bar{\omega}_j f_j = \bar{\omega}_j (p_j x + q_j y + r_j) \tag{4.3.5}$$

这里参数集$\{p_j, q_j, r_j\}$包括了结果参数。

第五层:聚合(和)层。

这一层的唯一节点是一个记为$\sum$的圆形节点,此层的输出计算了总输出量,即所有输入信号的总和,可表示为

$$O_{5,1} = \sum_j \bar{\omega}_j f_j = \frac{\sum_j \omega_j f_j}{\sum_j \omega_j} \tag{4.3.6}$$

4.4 禁忌搜索

一些算法可以用来在优化问题中寻找全局最小值,其中,禁忌搜索(TS)、遗传算法(GA)和 PSO 是很常见的演化算法。由 F. Glover 提出的禁忌搜索算法[18-19]被用来解决多目标组合的优化问题。此概念被扩展之后用来解决连续优化问题,比如连续禁忌搜索(CTS),该算法由 P. Siarry 和 G. Berthiau 提出[20]。增强连续禁忌搜索(ECTS)是一个使用了禁忌搜索中一些先进概念的算法,例如对连续变量的优化函数进行多样化和强化处理。ECTS 由 R. Chelouh 和 P. Siarry 提出[21]。

4.4.1 禁忌概念

在介绍 ECTS 算法之前需要解释 TS 中的定义和概念,图 4.5 显示了一系列在搜索过程中可能用到的解决办法,这被定义为搜索空间[22]。

搜索空间的维度等于代价函数里面变量的数目。搜索空间定义为X,搜索空间内的任意元素定义为$\boldsymbol{x}^i$,在n个变量的函数中,周围搜索空间是$\Re^n$,第i个元素为$\boldsymbol{x}^i = (x_1^i, x_2^i, \cdots, x_n^i)$。

元素$\boldsymbol{x}^i$的代价函数或者目标函数表示为$f(\boldsymbol{x}^i)$,对于一个有n个变量的函

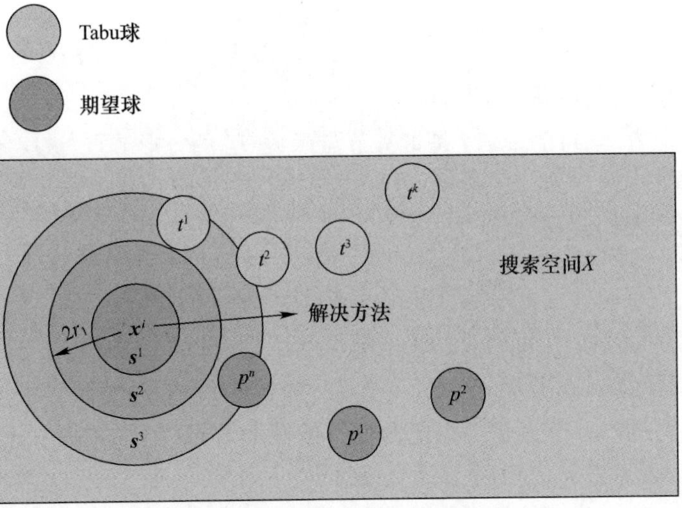

图 4.5　用于增强连续禁忌搜索的搜索空间的表征

数，$f(\boldsymbol{x}^i)$ 是 $\Re^n \to \Re^1$ 的映射关系。函数的目标是使下式最小化：

$$f(\boldsymbol{x}^i):\boldsymbol{x}^i \in X \tag{4.4.1}$$

对于空间 X 内的每个点都可以产生一系列的近邻，这样在搜索空间 X 便组成了一个子空间邻域空间，记作 $S \subset X$，每一个在 S 中的元素 j 记作 $\boldsymbol{s}^j=(s_1^j, s_2^j, \cdots, s_n^j)$，其中 n 是搜索空间 X 的维度或者变量数量。

在禁忌搜索（TL）里最重要的部分是禁忌列表（TL）。为了防止在搜索空间 X 中产生重复动作（循环），将使用一个称为禁忌列表的数组。禁忌列表记录了最近的运动，以便到过的位置在将来不会再次被访问。在禁忌列表里被记录的运动称为"Tabu"，它受到禁忌的激发，意思是对被一个群体、文化、社会或社区认为不可取或令人反感的语言、物品、行动或讨论的强烈社会禁令/禁令。Tabu 的迭代次数称作禁忌长度（TT），由禁忌列表的长度决定。数字 N_1 越大，禁忌长度越长，N_1 在禁忌搜索算法的实现过程中扮演重要角色，禁忌列表中的每个元素 k 表示为 $\boldsymbol{t}^k=(t_1^k, t_2^k, \cdots, t_n^k)$。

4.4.2　增强连续禁忌搜索

图 4.6 展示了 ECTS 的流程图，ECTS 使用了较 TS 更先进的概念，引入了多样化和强化。ECTS 由四个阶段组成：初始化参数，多样化，选择最理想区域以及强化。下文将对每一步都做详细介绍。

4.4.2.1　参数的初始化

在 ECTS 算法中需要进行初始化的参数有：TS 的长度（N_1），期望列表（PL）

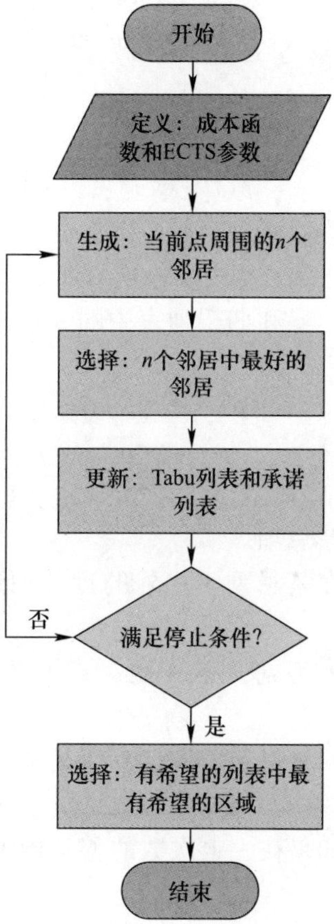

图 4.6　增强连续禁忌搜索流程图

的长度(N_2),邻域的半径 r_1,禁忌圆球的半径 r_2,期望圆球的半径 r_3,一个搜索空间 X 的随机点。

4.4.2.2　多样化

在这个阶段,算法在搜索空间 X 中寻找最有希望的区域,详细步骤如下:

(1) 均匀生成近邻。如图4.5所示,对于任意点 $\boldsymbol{x}^i \subset X$,$\boldsymbol{x}^i$ 生成 N 个近邻 $\boldsymbol{s}^j$,例如:

$$(j-1)r_1 \leqslant \|\boldsymbol{x}^{\mathrm{i}}-\boldsymbol{s}^j\|_2 \leqslant jr_1 \tag{4.4.2}$$

式中:$j=1,2,\cdots,N$;r_1 是邻域的初始半径。

$$\|\boldsymbol{x}_i-\boldsymbol{s}^j\|_2=\sqrt{(x_1^i-s_1^j)^2+\cdots+(x_n^i-s_n^j)^2} \tag{4.4.3}$$

上面的方法是将搜索区域划分成为同心球。

(2) 与禁忌列表进行比较。在与禁忌列表的每个元素 $\boldsymbol{t}^k(k=1,2,\cdots,N_1)$ 比较之前,会生成每个近邻 $\boldsymbol{s}^j(j=1,2,\cdots,N)$,如果

$$\|\boldsymbol{t}^k-\boldsymbol{s}^j\|_2\leqslant r_2 \tag{4.4.4}$$

式中

$$\|\boldsymbol{t}^k-\boldsymbol{s}^j\|_2=\sqrt{(t_1^k-s_1^j)^2+\cdots+(t_n^k-s_n^j)^2} \tag{4.4.5}$$

则相应的 $\boldsymbol{s}^j$ 被拒绝而成为禁忌项。

(3) 与期望列表进行比较。N' 表示没有在禁忌列表的近邻个数,并且记作 $\boldsymbol{h}^m(m=1,2,\cdots,N')$,这些元素和期望列表中的元素 $\boldsymbol{p}^n(n=1,2,\cdots,N_2)$ 进行比较,并且如果

$$\|\boldsymbol{p}^n-\boldsymbol{h}^m\|_2\leqslant r_3 \tag{4.4.6}$$

式中

$$\|\boldsymbol{p}^n-\boldsymbol{h}^m\|_2=\sqrt{(p_1^m-h_1^m)^2+\cdots+(p_n^n-h_n^m)^2} \tag{4.4.7}$$

则相应的 $\boldsymbol{s}^j$ 被拒绝而成为禁忌项。

(4) 寻找最佳近邻。在禁忌列表之外的近邻中选择代价函数最小的最佳近邻。

(5) 更新禁忌列表。在禁忌列表中更新得到的最佳近邻,采用先入先出(FIFO)方式。

(6) 更新期望列表。如果得到的最佳近邻是全局最优的,则将其更新至期望列表,采用先入先出方式。

(7) 转换确定条件。如果在一定次数的移动内没有得到提升,多样化过程被终止。

4.4.2.3 选择最理想区域

从期望列表里选择最理想区域有两种方法:恒定半径和标准差。

(1) 恒定半径:期望区域的上边界($\mathrm{PA_{ub}}$)、下边界($\mathrm{PA_{lb}}$)的定义是期望列表加减恒定半径 r_b 后再除以维度 n,可以表示为

$$\mathrm{PA_{ub}}=\frac{\mathrm{PL}+r_\mathrm{b}}{n} \tag{4.4.8}$$

$$\mathrm{PA_{lb}}=\frac{\mathrm{PL}-r_\mathrm{b}}{n} \tag{4.4.9}$$

(2) 标准差:期望区域的上边界和下边界可以由期望列表加减通过固定数量模拟得到的方差(SD),然后除以维度 n,可以表示为

$$\mathrm{PA_{ub}}=\frac{\mathrm{PL}+\mathrm{SD}}{n} \tag{4.4.10}$$

$$\mathrm{PA_{lb}}=\frac{\mathrm{PL}-\mathrm{SD}}{n} \tag{4.4.11}$$

4.4.2.4 强化

从期望列表中选择出最理想区域之后，通过重复步骤(1)～(7)在最理想区域内强化搜索。

4.5 遗传算法

1859年，达尔文(1809—1882)出版了《物种起源》，他认为所有生物都有很大的潜能去产生后代，比如蛋，但是只有很小一部分能生长到成年，面对随机多变的自然环境，生存着的物种为了生存不得不改变他们的特征。因此，进化是遗传变异的自然选择。

在同一时期，Gregor Mendel(1822—1884)在豌豆植物的实验中研究了遗传的特征。这些实验证明了达尔文的理论。此外，在Mendel去世三十年之后，Walter Sutton(1877—1916)发现了果蝇的基因是细胞核中染色体的一部分，这表明如果性状由单个基因控制，那么基因突变会产生巨大的影响，但是如果性状是一些基因结合起来控制，那么其中的一个发生变异产生的影响微乎其微。

遗传算法(GA)的灵感来源于为了适应随机多变的环境不得不改变性状的物种的染色体。因此，GA是一个基于自然生物进化隐喻的随机搜索和优化方法，它由一些运算符表示，比如选择、交叉、变异等。GA应用了适应、繁殖和变异的生存原理，并产生良好的近似解，所以GA被用来解决组合优化问题。GA的概念被扩展到解决连续优化问题，产生的算法如由P. Siarry和G. Berthiau提出的连续化遗传算法(CGA)[23]。CGA和GA相似，除了CGA的参数适用于连续数字，而GA的参数是用二进制格式编码的。CGA使用线性排序模型进行染色体选择，其中概率密度函数是基于单个候选染色体的代价函数生成的。

4.5.1 基础遗传算法过程

图4.7展示了遗传算法的流程图[24]，过程如下：

(1) 定义CGA参数：包括初始种群(Ipop)，第一次遗传结束的种群(pop)，用于交配的染色体数目(Keep)，突变率(Mut)，容忍值ε等。

(2) 生成均匀的种群：生成N个元素(染色体)，并且N是初始种群。

(3) 求每个染色体的代价函数：计算在种群中第i个成员的适合度值f_i。

(4) 根据每个基因的表现选择配对：根据当前适合度的排序，从当前种群中生成一个新的种群，比如，决定哪对父母适合在下次遗传过程中产生后代。

(5) 通过交叉繁殖产生后代：利用单个或者多个交叉点产生新的染色体，保持良好的特征，抛弃不好的特征。

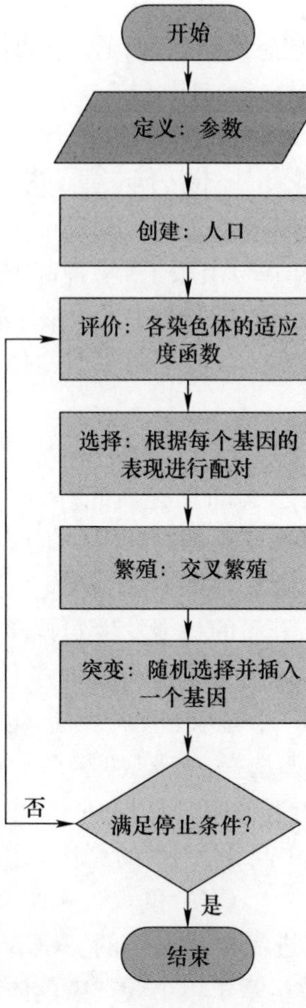

图 4.7　遗传算法流程图

(6) 突变:利用突变率对基因产生随机突变,防止陷入局部最优解。

(7) 重复步骤(3) ~ (6),直到达到迭代次数的最大值,或者满足 ε 条件停止。

4.6　粒子群算法

1995 年,一位社会心理学家 J. Kennedy 和 R. Eberhart 发现了一种新的进化算法——粒子群优化(PSO)[25]。他们从鸟群和鱼群的行为中获得数学运算启发[26],PSO 算法初始化的过程和 GA 相似,都是利用一个随机的种群,但 GA 在每代使用交叉和突变方式更新染色体。与 GA 不同的是,PSO 采用了粒子群的速度、局部最优位置和全局最优位置解法来更新解[25-30]。相比较于其他进化最优解法,PSO 算法使用几个简单的规则处理复杂的行为。所以在内存需求方面它的计算成本很低,并且耗时比 GA 少。很多研究人员已经在不同问题情境下成功证明了 PSO 算法的益处[31-45]。我们首先提供基础 PSO 的工作流程,包括步骤和公式,然后通过 5 种不同的技术和均匀、随机分布来研究 PSO 动态过程[46]。

2003 年,PSO 研究人员将 PSO 算法大体上分为 5 类:算法、拓扑学、参数、和其他进化计算模型一同出现以及应用[47]。在这些分类中,参数是最重要的。Clerc 和 Kennedy 通过收缩因素对膨胀、稳定性和收敛性进行了研究[36]。Shi 和 Eberhart 提出了惯性权重算法并且比较了两种方式:收缩因子和惯性权重算法[34],他们也研究了 PSO 算法中的参数选择[33]。然而,利用几个参数去保留算法的数值稳定性和准确性是最重要的。因此,提出了 PSO 的动态研究。

4.6.1 基础 PSO 的步骤和公式

PSO 有着过程简单的特性,如图 4.8 所示[46]。

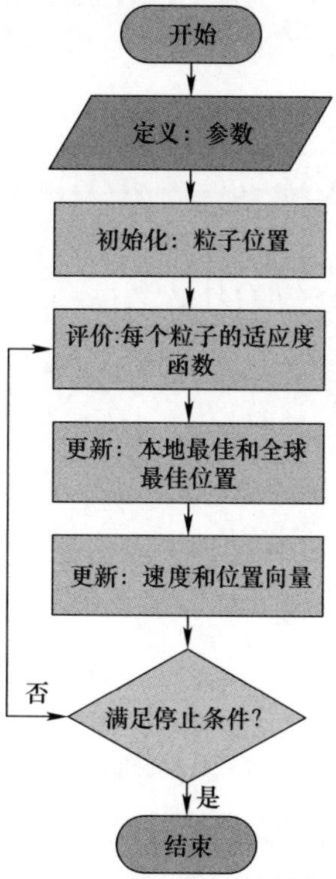

图 4.8 粒子群算法的流程图

其过程描述如下：

(1) 定义 PSO 的输入参数：包括迭代次数、群大小、速度极限($V_{\max}$)、搜索空间的上边界(hi)和下边界(lo)位置，终止容忍度 $\varepsilon > 0$。

(2) 初始化粒子位置：该过程用均匀分布或者正态分布来随机完成[46]。

(3) 评估第 j 个粒子的适合度函数品质：如果新的粒子位置能产生更优的代价函数，就用新的位置取代原来的位置，比如，更新局部最优和全局最优的代价和位置值。此外，考虑容忍值 ε 并且比较当前 t 和上次 $(t-1)$ 最优代价值的差值 $|f^*(t)-f^*(t-1)|$，如果 $|f^*(t)-f^*(t-1)|<\varepsilon$，则停止循环，否则进入下一步。

(4) 如图 4.9 所示，根据运动方程更新速度 $V_i^j(t)$ 和位置 $x_i^j(t)$ 矢量，有

$$V_i^j(t)=\alpha(t)V_i^j(t-1)+\beta(t)[x_i^{j,\text{lbest}}(t-1)-x_i^j(t-1)]+\gamma(t)[x_i^{j,\text{gbest}}(t-1)-x_i^j(t-1)] \tag{4.6.1}$$

$$x_i^j(t)=x_i^j(t-1)+V_i^j(t)\Delta t \tag{4.6.2}$$

式中：$V_i^j(t)$ 和 $V_i^j(t-1)$ 是时间分别为 t 和 $t-1$ 时，第 j 个粒子的第 i 部分的速度；$\alpha(t)$、$\beta(t)$、$\gamma(t)$ 是从均匀分布或者正态分布提取到的随机值，详细部分在 4.6.2 节解释；lbest 和 gbest 表示在每代的局部最优位置和全局最优位置；Δt 是每次迭代的间隔时间，比如，每次迭代 $\Delta t=1$，式(4.6.1)和式(4.6.2)可表示成如下矩阵：

$$\begin{bmatrix} V_i^j(t) \\ x_i^j(t) \end{bmatrix}=\begin{bmatrix} \alpha(t) & -\beta(t)-\gamma(t) \\ \alpha(t) & 1-\beta(t)-\gamma(t) \end{bmatrix}\begin{bmatrix} V_i^j(t-1) \\ x_i^j(t-1) \end{bmatrix}+\begin{bmatrix} \beta(t) & \gamma(t) \\ \beta(t) & \gamma(t) \end{bmatrix}\begin{bmatrix} x_i^{j,\text{lbest}}(t-1) \\ x_i^{j,\text{gbest}}(t-1) \end{bmatrix} \tag{4.6.3}$$

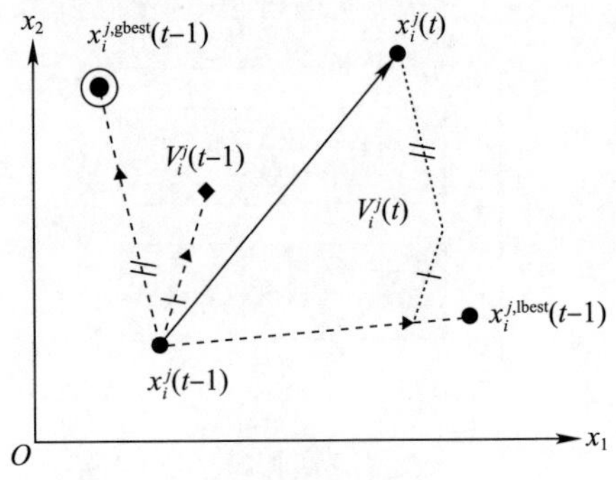

图 4.9 粒子群前一刻和更新后的位置图

需要注意的是，如果粒子的速度值过大，他们会使粒子群经常逃离搜索空间[26-28]。因此，为了避免粒子群的膨胀，需要一个速度上限值 $V_{\max}$[26]。例如：如果速度值 $V_i^j(t)>V_{\max}$，那么 $V_i^j(t)=V_{\max}$；如果 $V_i^j(t)<-V_{\max}$，那么 $V_i^j(t)=-V_{\max}$。

同样的,边界情况也应该考虑,比如:如果更新的位置 $x_i^j(t)>\mathrm{hi}$,那么 $x_i^j(t)=\mathrm{hi}$;如果更新的位置 $x_i^j(t)<\mathrm{lo}$,那么 $x_i^j(t)=\mathrm{lo}$,这里下标 i 是粒子维度的组成,上标 j 代表粒子群的序号,t 代表时间。

(5) 步骤(2)~(4)直到达到迭代次数的最大值,或者满足 ε 条件停止。

4.6.2 五种不同的 PSO 技术

根据式(4.6.1)~式(4.6.3),每次迭代之后基于局部最优位置和全局最优位置原则更新速度和位置,下面是根据局部最优和全局最优来决定下一个速度和位置的五种技术。

(1) 群局部最优。在这种技术中,每次更新的局部最优代价由每次迭代过程中的粒子局部最优位置表示[26],如图 4.10 所示。

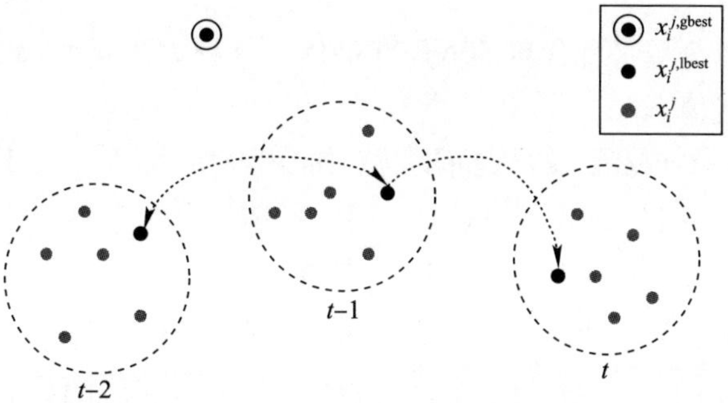

图 4.10 群局部最优的每次运动图示

(2) 粒子的局部最优。图 4.11 和图 4.12 对比了在两种不同局部最优技术的适合度代价[28]。

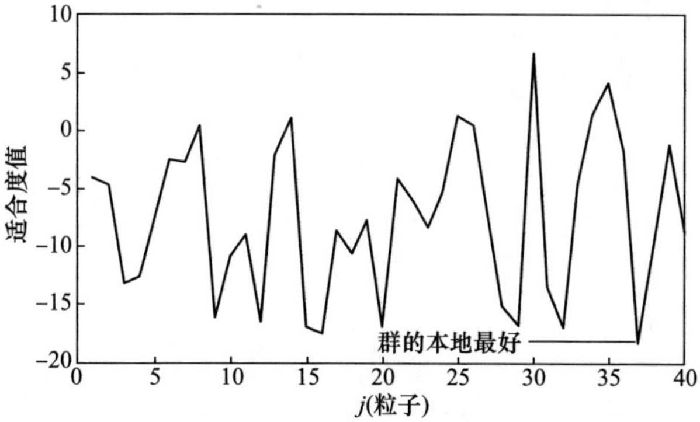

图 4.11 群局部最优图示

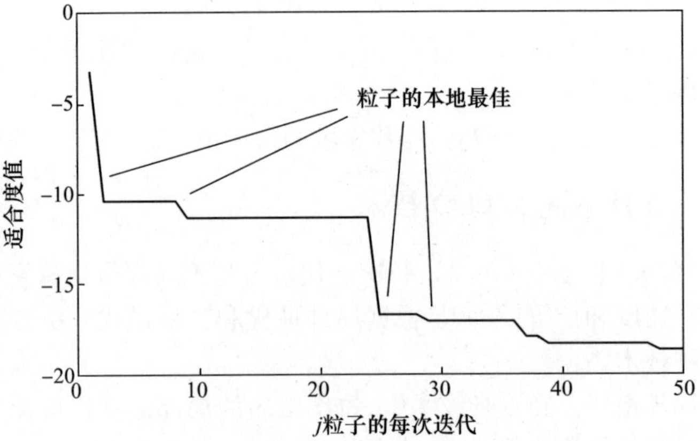

图 4.12　粒子局部最优图示

(3) 不带有全局最优代价的群局部最优。这种方法是技术 A 的简化,只是忽略了全局最优值。

(4) 带有样条加权函数的群局部最优。如图4.9 所示,由 $V_i^j(t-1)$,$x_i^{j,\mathrm{lbest}}(t-1)-x_i^j(t-1)$和 $x_i^{j,\mathrm{gbest}}(t-1)-x_i^j(t-1)$的增加量乘一个随机值决定,然而,如果粒子 j 离群局部最优或者全局最优位置过近,则 $x_i^{j,\mathrm{lbest}}(t-1)-x_i^j(t-1)$ 和 $x_i^{j,\mathrm{gbest}}(t-1)-x_i^j(t-1)$的影响会变得很小。相反,如果粒子 j 离群局部最优或者全局最优位置过远,则 $x_i^{j,\mathrm{lbest}}(t-1)-x_i^j(t-1)$ 和 $x_i^{j,\mathrm{gbest}}(t-1)-x_i^j(t-1)$ 的影响应该增加。考虑到速度连续性和数值稳定性,选用一个样条加权函数给一个权重函数 $W_i^{j,\mathrm{lbest}}(t-1)$ 和 $W_i^{j,\mathrm{gbest}}(t-1)$ 建模[48],给出如下等式:

$$W_i^{j,\mathrm{lbest}}=\begin{cases}6\left(\dfrac{d_i^{j,\mathrm{lbest}}}{r_i}\right)^2-8\left(\dfrac{d_i^{j,\mathrm{lbest}}}{r_i}\right)^3+3\left(\dfrac{d_i^{j,\mathrm{lbest}}}{r_i}\right)^4, & 0\leqslant d_i^{j,\mathrm{lbest}}\leqslant r_i\\ 1, & r_i\leqslant d_i^{j,\mathrm{lbest}}\end{cases}$$

和

$$W_i^{j,\mathrm{gbest}}=\begin{cases}6\left(\dfrac{d_i^{j,\mathrm{gbest}}}{r_i}\right)^2-8\left(\dfrac{d_i^{j,\mathrm{gbest}}}{r_i}\right)^3+3\left(\dfrac{d_i^{j,\mathrm{gbest}}}{r_i}\right)^4, & 0\leqslant d_i^{j,\mathrm{gbest}}\leqslant r_i\\ 1, & r_i\leqslant d_i^{j,\mathrm{gbest}}\end{cases}$$

式中:$d_i^{j,\mathrm{lbest}}=|x_i^j-x_i^{j,\mathrm{lbest}}|$;$d_i^{j,\mathrm{gbest}}=|x_i^j-x_i^{j,\mathrm{gbest}}|$;$d_i^{j,\mathrm{lbest}}$ 是第 j 个粒子与局部最优位置 $x_i^{j,\mathrm{lbest}}$ 之间的距离;$d_i^{j,\mathrm{gbest}}$ 是第 j 个粒子与全局最优位置 $x_i^{j,\mathrm{gbest}}$ 之间的距离;r_i 是支持域的半径,如图 4.13 所示。因此式(4.6.1)和式(4.6.3)可以改写为

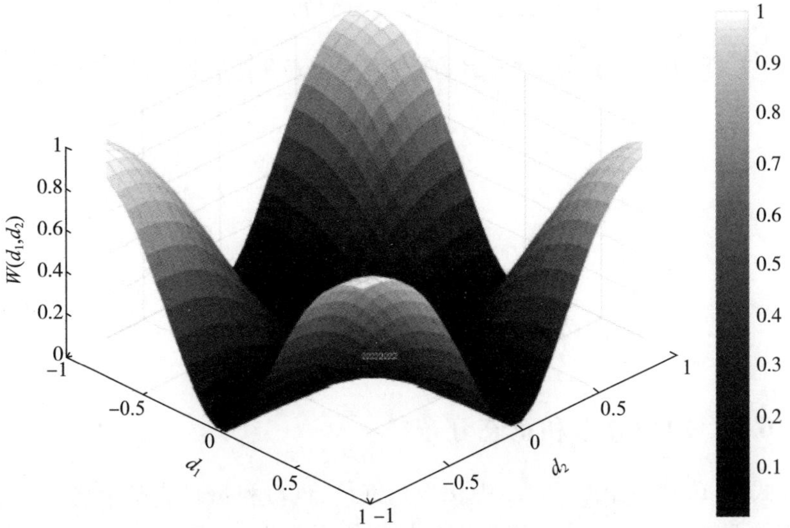

图 4.13　样条加权函数

$$V_i^j(t)=\alpha(t)V_i^j(t-1)+W_i^{j,\mathrm{lbest}}\beta(t)(x_i^{j,\mathrm{lbest}}(t-1)-x_i^j(t-1))+$$
$$W_i^{j,\mathrm{gbest}}\gamma(t)(x_i^{j,\mathrm{gbest}}(t-1)-x_i^j(t-1)) \tag{4.6.4}$$

和

$$\begin{bmatrix} V_i^j(t) \\ x_i^j(t) \end{bmatrix}=\begin{bmatrix} \alpha(t) & -\beta(t)W_i^{j,\mathrm{gbest}}-\gamma(t)W_i^{j,\mathrm{gbest}} \\ \alpha(t) & 1-\beta(t)W_i^{j,\mathrm{gbest}}-\gamma(t)W_i^{j,\mathrm{gbest}} \end{bmatrix}\begin{bmatrix} V_i^j(t-1) \\ x_i^j(t-1) \end{bmatrix}+$$
$$\begin{bmatrix} \beta(t)W_i^{j,\mathrm{lbest}} & \gamma(t)W_i^{j,\mathrm{gbest}} \\ \beta(t)W_i^{j,\mathrm{lbest}} & \gamma(t)W_i^{j,\mathrm{gbest}} \end{bmatrix}\begin{bmatrix} x_i^{j,\mathrm{lbest}}(t-1) \\ x_i^{j,\mathrm{gbest}}(t-1) \end{bmatrix} \tag{4.6.5}$$

（5）带有样条加权函数和适合度函数梯度f的群局部最优。共轭梯度的技术具有 GA 收敛快的特性[49]。我们介绍一个改进型下降法模拟共轭梯度。在带有样条加权技术的模拟梯度算法中,待更新位置,式(4.6.2),被写为

$$x_i^j(t)=x_i^j(t-1)+V_i^j(t)\Delta t-\delta(t)W_i^{j,\mathrm{gbest}}\frac{\nabla_i f}{\|f\|_2} \tag{4.6.6}$$

式中:$\delta(t)$是一个随机变量,在 4.6.3 节详细阐述; $\|f\|_2$ 是适合度函数f的范数;$\nabla_i f$是在方向i上适合度函数f的梯度。式(4.6.3)也可以写为

$$\begin{bmatrix} V_i^j(t) \\ x_i^j(t) \end{bmatrix} = \begin{bmatrix} \alpha(t) & -\beta(t)W_i^{j,\text{lbest}} - \gamma(t)W_i^{j,\text{gbest}} \\ \alpha(t) & 1-\beta(t)W_i^{j,\text{lbest}} - \gamma(t)W_i^{j,\text{gbest}} \end{bmatrix} \begin{bmatrix} V_i^j(t-1) \\ x_i^j(t-1) \end{bmatrix} +$$

$$\begin{bmatrix} \beta(t)W_i^{j,\text{lbest}} & \gamma(t)W_i^{j,\text{gbest}} \\ \beta(t)W_i^{j,\text{lbest}} & \gamma(t)W_i^{j,\text{gbest}} \end{bmatrix} \begin{bmatrix} x_i^{j,\text{lbest}}(t-1) \\ x_i^{j,\text{gbest}}(t-1) \end{bmatrix} -$$

$$\begin{bmatrix} 0 \\ \delta(t)W_i^{j,\text{gbest}} \dfrac{\nabla_i f}{\|f\|_2} \end{bmatrix} \tag{4.6.7}$$

4.6.3 均匀分布和正态分布

在式(4.6.1)~式(4.6.7)中,$\alpha(t)$、$\beta(t)$、$\gamma(t)$和$\delta(t)$是基于均匀分布或正态分布的随机变量。基于均匀分布变量值用 Uni 表示,且 Uni $\in[0,1]$,而基于正态分布的变量值用$N(\mu,\sigma)$表示,其中μ是平均值,σ是标准差,通常其概率密度函数如下:

$$f(x) = \frac{\exp\left(\dfrac{-1}{2}\left(\dfrac{x-\mu}{\sigma}\right)^2\right)}{\sigma\sqrt{2\pi}} \tag{4.6.8}$$

图4.14表示了正态分布,这里$N(0,1)$、$N(1,2)$、$N(1,1)$、$N(1,1/2)$、$N(1,1/3)$分别由 Nor1、Nor2a、Nor2b、Nor2c、Nor2d表示。

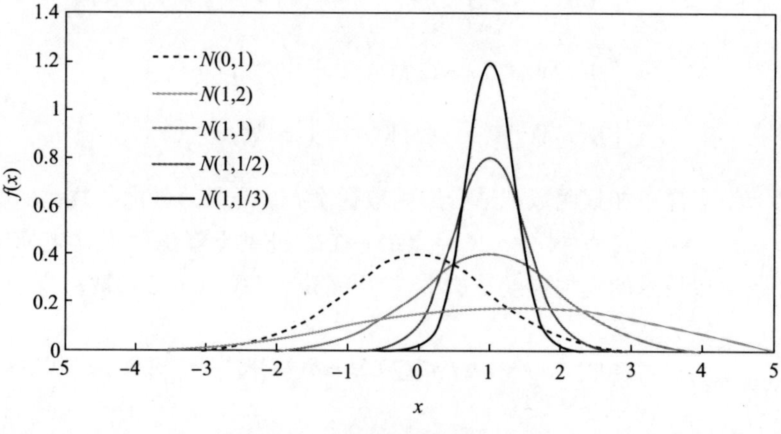

图4.14 单个正态分布图示

4.7 自适应粒子群优化

4.6.1 中提到，在基础 PSO 算法的步骤 4 里面，最大速度值的设定是为了避免粒子膨胀，然而对于多维度问题，V_{max} 应该能够根据情况进行自适应调节，使系统工作性能更优。因此，我们提出了自适应 PSO（APSO）算法去改变 V_{max}[50]。

4.7.1 APSO 过程和公式

APSO 首先建立了线性模型。从粒子群中选择 $n+1$ 个点，然后选取群中剩余点计算相对误差标准差。最后，使用基于计算相对误差标准偏差的模糊逻辑规则和展开的线性平面倾斜度来适应速度的上限 V_{max}，详细过程如下：

（1）构造一个线性模型。图 4.15 是在 n 维空间中群的位置和代价示意图，$\Re^n$ 中的矢量是 r，$\Re^{n+1}$ 中的矢量是 $\boldsymbol{R}$，所以第 j 个粒子的位置矢量 $\boldsymbol{r}_j$ 和 $\boldsymbol{R}_j$ 可以表示为

$$\boldsymbol{r}_j = (x_1^j(t), x_2^j(t), \cdots, x_i^j(t)) \tag{4.7.1}$$

$$\boldsymbol{R}_j = (\boldsymbol{r}_j, c_j) \tag{4.7.2}$$

式中：$j=1,2,\cdots,S_s$，S_s 是群大小；$i=n$ 表示多维空间的大小；c_j 表示第 j 个粒子的代价函数。

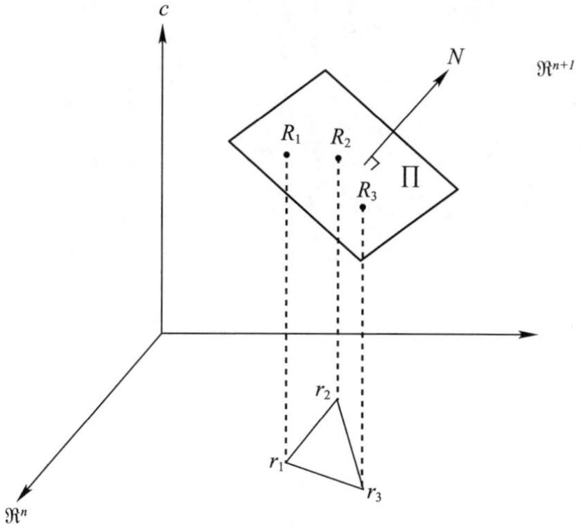

图 4.15　群的位置和代价在 n 维空间示意图

通过在群大小 S_s 中随机挑选 $n+1$ 个粒子，所得适合这些点的平面如下：

$$\boldsymbol{N}\cdot\boldsymbol{R}=b \tag{4.7.3}$$

式中：$\boldsymbol{N}=(N_1,N_2,\cdots N_{n+1})$代表所产生平面$\prod$的法矢量，$N_1,N_2,\cdots,N_{n+1}$是$\boldsymbol{N}$的非零量，注意$\boldsymbol{N}$和$b$没有唯一的定义（两者的任意倍数描述相同的平面），但是必须保证线性模型的“平面”永远不会垂直。因此，需要令$N_{n+1}=1$，式（4.7.3）可写为

$$\boldsymbol{R}\cdot\boldsymbol{N}-b=0 \tag{4.7.4}$$

在$N_{n+1}=1$的条件下对于$\boldsymbol{R}=\boldsymbol{R}_1,\boldsymbol{R}_2,\cdots,\boldsymbol{R}_{n+1}$，将（4.7.4）改写成矩阵形式：

$$\boldsymbol{A}\boldsymbol{u}=\boldsymbol{w} \tag{4.7.5}$$

式中

$$\boldsymbol{A}=\begin{bmatrix} r_1 & c_1 & -1 \\ r_2 & c_2 & -1 \\ \vdots & \vdots & \vdots \\ r_{n+1} & c_{n+1} & -1 \\ 0 & 1 & 0 \end{bmatrix} \tag{4.7.6}$$

$$\boldsymbol{u}=\begin{bmatrix} N_1 \\ N_2 \\ \vdots \\ N_{n+1} \\ b \end{bmatrix}=\begin{bmatrix} N_1 \\ N_2 \\ \vdots \\ 1 \\ b \end{bmatrix},\quad \boldsymbol{w}=\begin{bmatrix} 0 \\ 0 \\ \vdots \\ 0 \\ 1 \end{bmatrix} \tag{4.7.7}$$

这里$\boldsymbol{u}$由$\boldsymbol{u}=\boldsymbol{A}^{-1}\boldsymbol{w}$解出。现在得到描述代价曲面的线性方程，这些代价曲面由$\boldsymbol{r}_1,\cdots,\boldsymbol{r}_{n+1}$和$c_1,\cdots,c_{n+1}$的选择决定。代价平面近似于

$$\boldsymbol{N}\cdot\boldsymbol{R}=b$$

$$(N_1,\cdots,N_n,N_{n+1})\cdot\boldsymbol{R}=b$$

或者

$$(N_1,\cdots,N_n,1)\cdot(\boldsymbol{r},c)=b \tag{4.7.8}$$

式中：$\boldsymbol{r}=(x_1,x_2,\cdots,x_n)\in\Re^n$。然后求解$c$：

$$c=-(N_1,\cdots,N_n)\cdot\boldsymbol{r}+b=L(\boldsymbol{r}) \tag{4.7.9}$$

$L(\boldsymbol{r})$是为代价函数$f(\boldsymbol{r})$计算出来的线性模型，比如，$L(\boldsymbol{r})\approx f(\boldsymbol{r})$。

（2）计算相对误差标准差和倾斜度。用$L(\boldsymbol{r})$去计算偏差（代价平面中偏差

的线性拟合)和标准差。从群大小 S_s 中随机选取 k(由 $k=S_s-(n+1)$决定)个点并且比较下面的矢量:

$$\begin{cases}(\tilde{\boldsymbol{r}}_1,\tilde{c}_1)=:\tilde{\boldsymbol{R}}_1\\(\tilde{\boldsymbol{r}}_2,\tilde{c}_2)=:\tilde{\boldsymbol{R}}_2\\\quad\vdots\\(\tilde{\boldsymbol{r}}_k,\tilde{c}_k)=:\tilde{\boldsymbol{R}}_k\end{cases}\tag{4.7.10}$$

e_i 定义为

$$e_i:=\tilde{c}_i-L(\tilde{\boldsymbol{r}}_i)\tag{4.7.11}$$

式中:$i=1,2,\cdots,k$; $\tilde{c}_i$是粒子 i 的代价值。

把(4.7.9)代入(4.7.11)然后由(4.7.12)得到偏差 e_i:

$$\begin{aligned}e_i&=\tilde{c}_i-L(\tilde{\boldsymbol{r}}_i)\\&=(N_1,\cdots,N_n)\cdot\boldsymbol{r}_i+\tilde{c}_i-b\end{aligned}\tag{4.7.12}$$

偏差数组 $\boldsymbol{e}$ 由下式计算:

$$\boldsymbol{e}=\begin{bmatrix}e_1\\e_2\\\vdots\\e_{k-1}\\e_k\end{bmatrix}=\begin{bmatrix}\tilde{\boldsymbol{r}}_1&\tilde{c}_1&1\\\tilde{\boldsymbol{r}}_2&\tilde{c}_2&1\\\vdots&\vdots&\vdots\\\tilde{\boldsymbol{r}}_{k-1}&\tilde{c}_{k-1}&1\\\tilde{\boldsymbol{r}}_k&\tilde{c}_k&1\end{bmatrix}\begin{bmatrix}N_1\\\vdots\\N_n\\1\\-b\end{bmatrix}=[\tilde{\boldsymbol{R}}\quad\boldsymbol{1}]\tilde{\boldsymbol{u}}\tag{4.7.13}$$

式中

$$\tilde{\boldsymbol{R}}=\begin{bmatrix}\tilde{\boldsymbol{r}}_1&\tilde{c}_1\\\tilde{\boldsymbol{r}}_2&\tilde{c}_2\\\vdots&\vdots\\\tilde{\boldsymbol{r}}_{k-1}&\tilde{c}_{k-1}\\\tilde{\boldsymbol{r}}_k&\tilde{c}_k\end{bmatrix},\quad\tilde{\boldsymbol{u}}=\begin{bmatrix}N_1\\\vdots\\N_n\\1\\-b\end{bmatrix}\tag{4.7.14}$$

样例(选中的 k 个点)的误差标准差 S_k 为

$$S_k = \sqrt{\frac{1}{k-1}\sum_{i=1}^{k}(\tilde{c}_i - L(\tilde{\boldsymbol{r}}_i))^2}$$

$$= \sqrt{\frac{1}{k-1}\sum_{i=1}^{k} e_i^2} \tag{4.7.15}$$

群的代价标准差为 S_n,所以相对误差标准差 S_{rel} 为

$$S_{\mathrm{rel}} = \frac{S_k}{S_n} \tag{4.7.16}$$

接下来,可以计算线性平面 Π 的倾斜度,平面 Π 的归一化法矢量 $\boldsymbol{N}_{\mathrm{s}}$ 由下式计算:

$$\boldsymbol{N}_{\mathrm{s}} = \frac{|\boldsymbol{N}'|}{\|\boldsymbol{N}'\|_2} = \frac{|\boldsymbol{N}'|}{\sqrt{\boldsymbol{N}'\boldsymbol{N}}} \in [0,1] \tag{4.7.17}$$

式中:$\boldsymbol{N} = [N_1, N_2, \cdots N_{n+1}]$。“小”倾斜度代表 $\boldsymbol{N}_{\mathrm{s}}$ 接近 1,“大”倾斜度代表 $\boldsymbol{N}_{\mathrm{s}}$ 接近 0。

(3) 模糊逻辑规则和隶属函数。如图 4.16 所示,下面提出 4 种模糊逻辑规则。

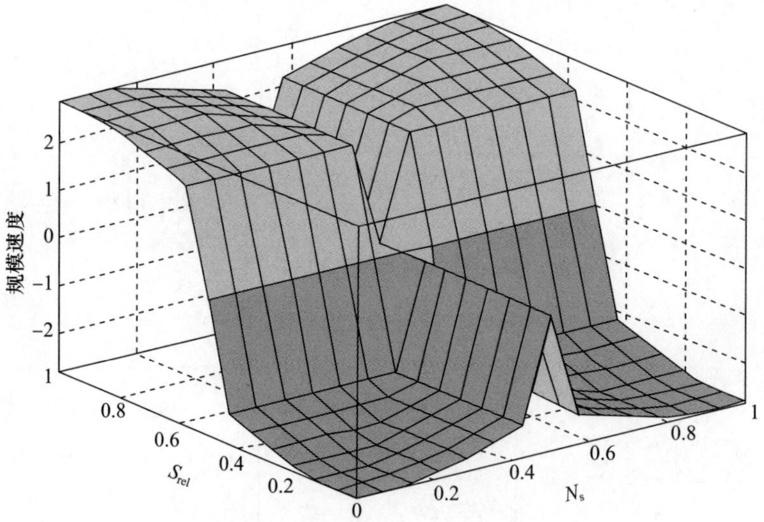

图 4.16 APSO 模糊逻辑规则平面

(1) 如果 $\boldsymbol{N}_{\mathrm{s}}$ 是“小”并且 S_{rel}是“大”,那么“减小”速度范围 S_v。

(2) 如果 $\boldsymbol{N}_{\mathrm{s}}$ 是“小”并且 S_{rel}是“小”,那么对比代价和上一次“平面”区域的代价,如果当前的代价小于之前平面区域的代价,那么就“增加”速度范围 S_v;然

而，如果当前的代价大于之前平面区域的代价，那么就“减小”速度范围 S_v。

(3) 如果 $\boldsymbol{N}_s$ 是“大”并且 S_{rel} 是“小”，那么“增大”速度范围 S_v。

(4) 如果 $\boldsymbol{N}_s$ 是“大”并且 S_{rel} 是“大”，那么“减小”速度范围 S_v。

这个过程中用到了两个隶属函数：Z 型和 S 型关于 x 的隶属函数 $f_z(x)$ 和 $f_s(x)$，表示如下：

$$f_z(x)=\begin{cases}1, & x\leqslant a_z\\ 1-2\left(\dfrac{x-a_z}{b_z-a_z}\right)^2, & a_z\leqslant x\leqslant\dfrac{a_z+b_z}{2}\\ 2\left(\dfrac{b_z-x}{b_z-a_z}\right)^2, & \dfrac{a_z+b_z}{2}\leqslant x\leqslant b_z\\ 0, & x\geqslant b_z\end{cases}$$

$$f_s(x)=\begin{cases}0, & x\leqslant a_s\\ 2\left(\dfrac{x-a_s}{b_s-a_s}\right)^2, & \dfrac{a_s+b_s}{2}\leqslant x\leqslant b_s\\ 1-2\left(\dfrac{b_s-x}{b_s-a_s}\right)^2, & a_s\leqslant x\leqslant\dfrac{a_s+b_s}{2}\\ 1, & x\geqslant b_s\end{cases}$$

式中：参数 a_z、b_z、a_s 和 b_s 从表 4.1 中选择出来。表中 $\boldsymbol{N}_s$ 和 S_{rel} 是输入变量，S_v 是输出变量。自适应速度最大值 V_{max} 由下式计算：

$$V_{max}=\exp(S_v) \tag{4.7.18}$$

表 4.1 隶属函数参数的选择

变量	变化	Z/S	a_z/a_s	b_z/b_s
$\boldsymbol{N}_s$	大	Z	0	0.5
$\boldsymbol{N}_s$	小	S	0.5	1
$\boldsymbol{S}_{rel}$	小	Z	0	0.5
$\boldsymbol{S}_{rel}$	大	S	0.5	1
S_v	增加	Z	−4	0
S_v	减少	S	0	4

4.7.2 改变/未改变的速度方向

在基础 PSO 过程(4.6.1 节)的步骤 4 中，V_{max} 约束更新速度，但是更新速度方向的变化如图 4.17(a) 所示。我们提出了一个不改变更新速度方向的方法,如图 4.17(b) 所示,更新速度的方向不因更新速度单位矢量和 V_{max} 而改变。

$$V(t) = \frac{V(t-1)}{\|V(t-1)\|_2} V_{max} \qquad (4.7.19)$$

(a) 改变更新速度方向   (b) 不改变更新速度方向

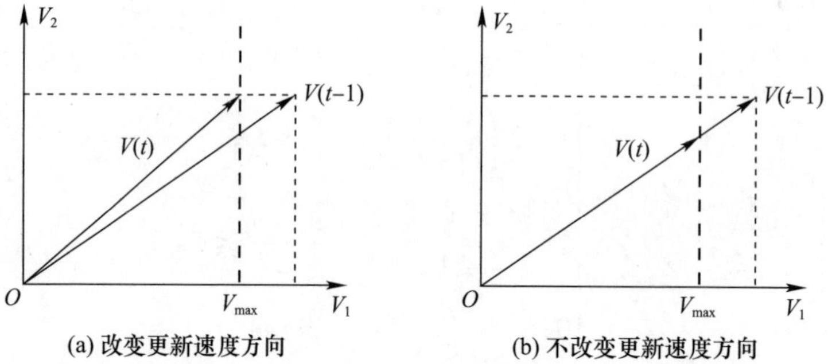

图 4.17　更新速度方法

4.8　凝聚混合优化

TS 和 PSO 各有优缺点,TS 能够在搜索空间中搜索广泛的区域,但是不能保证落入全局最优解,然而,实验结果表明,TS 确实找到了接近全局最优解的解决方案[51]。PSO 本身速度比较慢,但收敛性优于 TS,PSO 的性能可以通过一些额外的信息源得到加强,比如搜索空间。TS 和 PSO 的缺点可以通过组合它们得到弥补,并且二者的组合优化问题已经进行了尝试[52-53],我们以前的工作[54-55]提出了结合 ECTS 和 CGA 混合算法来解决有色噪声存在下的二维和三维连续优化问题和参数估计问题。尽管如此,GA 的实施还是比较复杂,需要大量的内存,并且消耗大量的 CPU 时间。

CHO 算法,包括了 ECTS 算法的多样性部分,和一个由 PSO 代替 ECTS 的集约化部分。换句话说,使用 PSO 算法加强了在搜索空间中对最有希望区域的搜索,CHO 算法的流程图如 4.18 所示,多样化和集约化分别来自 ECTS 和 PSO。

CHO 的多样化包含以下六个步骤。

(1) 定义:代价函数和 ECTS 参数。

(2) 生成:当前点周围 N 个近邻填入禁忌(Tabu)列表或者期望(Promising)列表。

(3) 选择:N 个近邻中最佳近邻,并且将它设置为新的当前点。

(4) 更新:禁忌列表和期望列表。

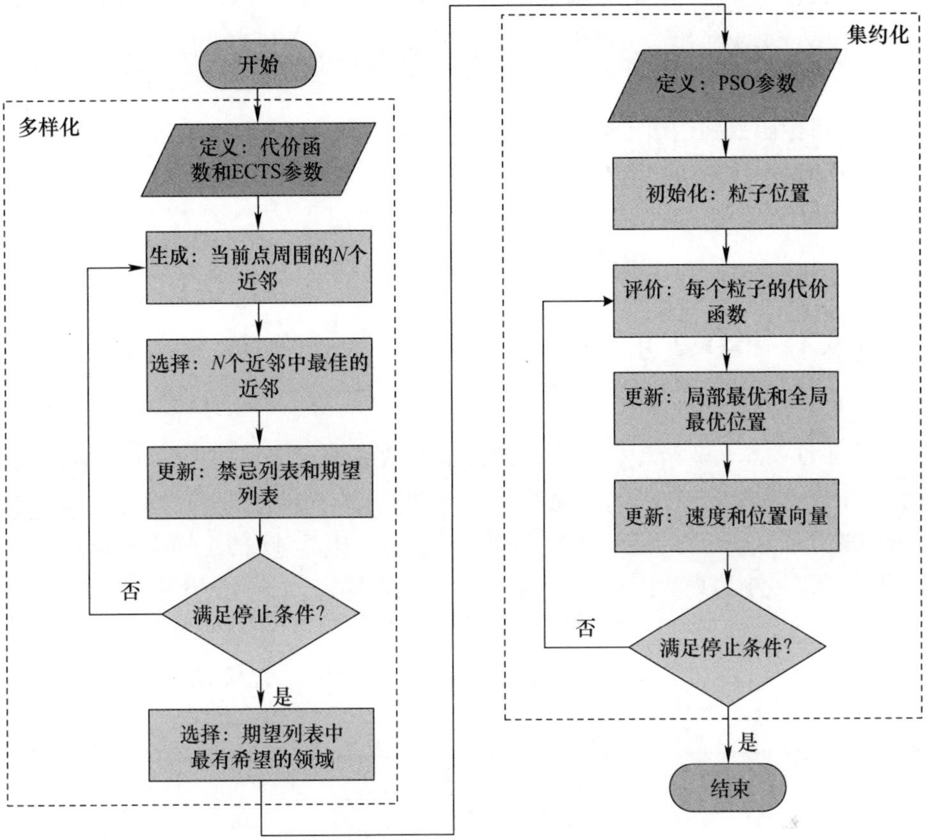

图 4.18　CHO 算法的多样化和集约化循环流程图

（5）停止：内循环在到达终止条件之一时停止。终止条件定义如下：

① 在达到固定的迭代次数时终止（最大迭代次数为 $I_{\max}$）。

② 经过多次迭代后，代价函数没有任何提升，比如，所有减少都小于容忍值 ε。

③ 目标函数到达预定值时。

（6）选择：从期望列表中选择最有希望的区域。

CHO 的集约化包含以下六个步骤：

（1）定义：代价函数和 PSO 参数。

（2）评估：每个粒子的代价函数。

（3）更新：局部最优和全局最优位置。

（4）更新：每个粒子的速度和位置矢量。

（5）停止：内循环在到达终止条件之一时停止。终止条件定义如下：

① 在达到固定的迭代次数时终止(最大迭代次数)

② 经过多次迭代后,代价函数没有任何提升,比如,所有减少都小于容忍值 ε。

③ 目标函数到达预定值时。

(6) 结束:完成 CHO 算法。

4.9 仿真结果及分析

4.9.1 PSO 动力学研究

4.9.1.1 基准问题

为了证明 4.6.2 节中提到的五种技术的精确性,我们对两个非线性问题进行了仿真。如图 4.19 所示,它是第一个问题 $f_1 = x_1\sin(4x_1) + 1.1x_2\sin(2x_2)$ 在 0 ~ 10的分配。在坐标为(9.0390,86682),值为 -18.5547 的全局最小值点附近有几个局部最小值点。图 4.20 和图 4.21 展示了第二个问题(Rosenbrock) $f_2 = 100(x_2 - x_1^2)^2 + (x_1 - 1)^2$ 分别在[-10,10]和[-1,1](注意坐标轴)范围内的分配。在点(1,1)处取得全局最小值 0,其他很多优化算法都很难找到这个点。

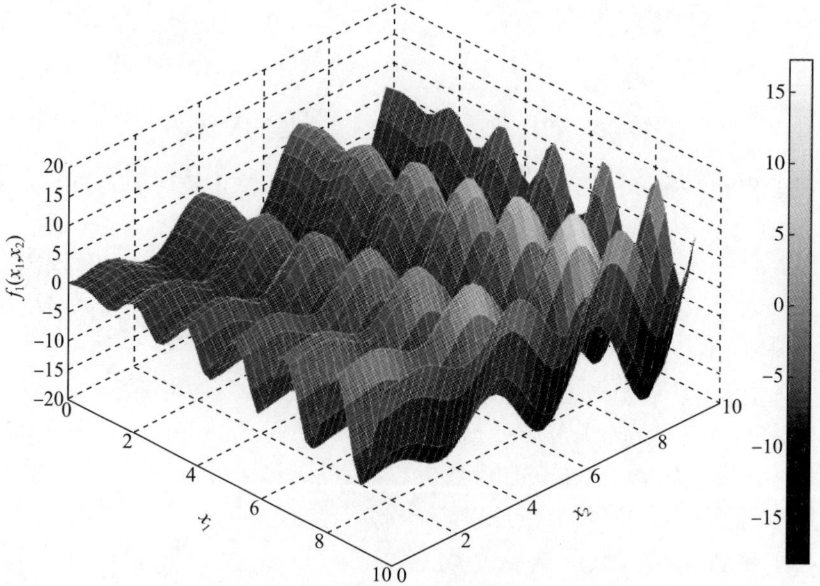

图 4.19 函数 f_1 在 0 ~ 10 之间的分布

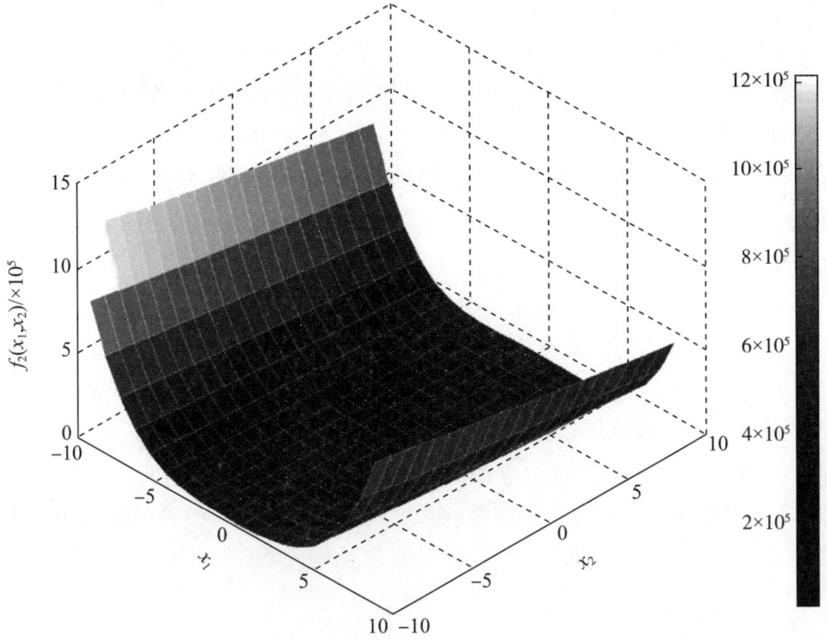

图 4.20　函数 f_2 在 0 ~ 10 之间的分布

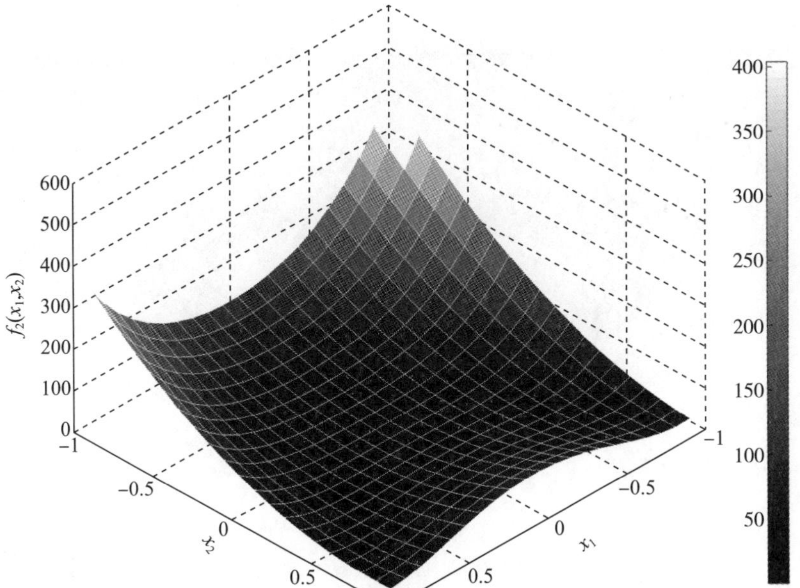

图 4.21　函数 f_2 在 −1 ~ 1 的分布

4.9.1.2 挑选参数

从4.6.1节的步骤一中可以看出，在执行步骤(2)～(5)时，对于初始参数的选择非常重要。因此，应当先选好三个参数最大迭代次数，群体大小和最大限制速度 V_{max} 跟样条支持域 r_i 半径的值再进行4.6.1节中的五个仿真实验。从仿真结果来看，第一个参数和最大迭代次数在两个问题中都被设置成了50。在两个问题中，第四个参数 r_i 的值都应当设置在8.5～15.0之间。而我们对所有的情况都设置成了 $r_i=10$。

对于两个仿真问题，一共有24种不同的情况对应，技术A或B和不同的随机分配(Uni、Nor1、Nor2a、Nor2b、Nor2c 或 Nor2d)。例如，图4.22是第一个问题适应度值f_1 在不同群体大小和最大速度下，在技术A和均匀分布下的分布；图4.23是Rosenbrock问题适应度值f_2 随技术B和正态分布 $N(1,1/3)$ 在不同群体大小和最大速度下的分布。考虑到统计学的影响，每次的仿真结果平均迭代30次。因此，得到24张图。表4.2说明每种情况所选的参数，最大速度(V_{max})和群体大小。表4.3是根据表4.2的并集选择的参数。

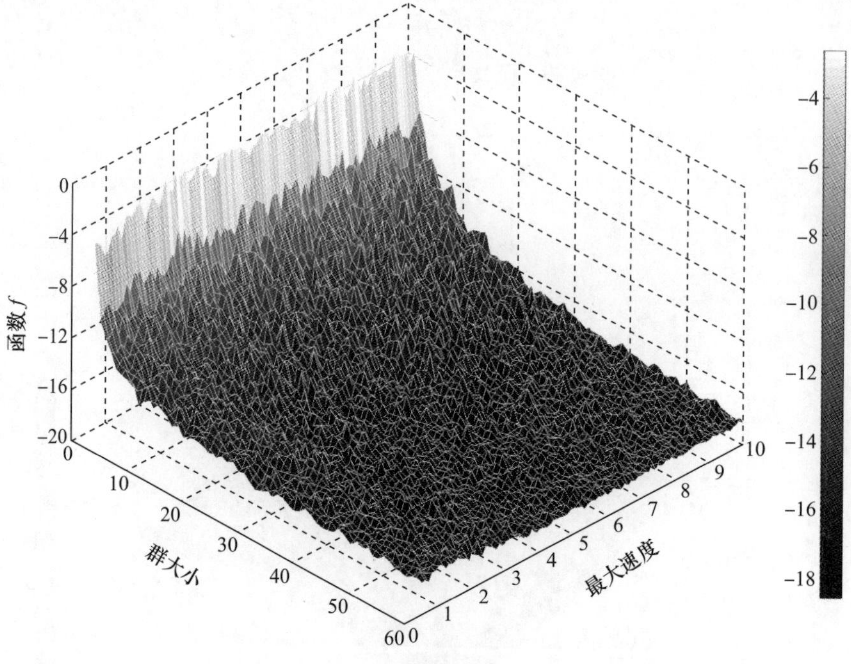

图4.22 不同群体大小和最大速度下技术A和Uni的函数f_1 分布

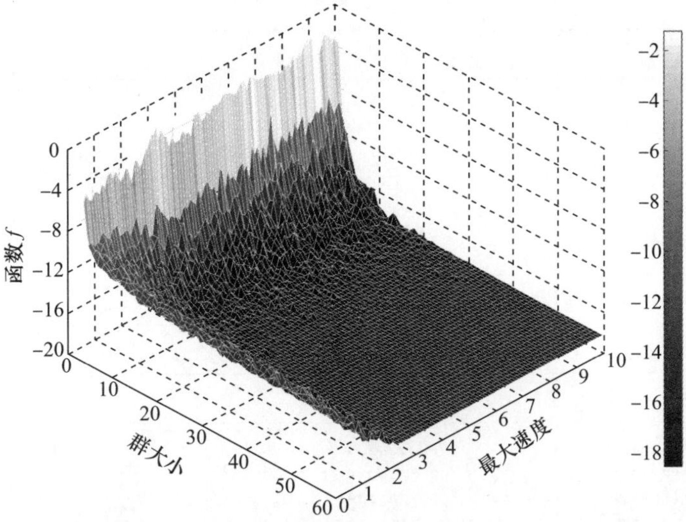

图 4.23　不同群体大小和最大速度下技术 B 和 Nor2d 的函数 f_2 分布

表 4.2　每种情况挑选的 V_{max} 和群大小

技术		问题 1		问题 2	
		V_{max}	群大小	V_{max}	群大小
A	Uni	[3.0,10.0]	35 +	[0.5,3.0]	40 +
	Nor1	[1.0,10.0]	25 +	[2.5,4.0]	60 +
	Nor2a	[0.5,3.0] 或 [8.5,9.2]	25 +	[0.5,3.0]	40 +
	Nor2b	[0.5,10.0]	25 +	[0.5,3.0]	30 +
	Nor2c	[2.0,10.0]	20 +	[0.2,3.5]	20 +
	Nor2d	[2.0,10.0]	20 +	[0.2,4.0]	20 +
B	Uni	[2.0,10.0]	35 +	[0.5,3.0]	30 +
	Nor1	[0.5,10.0]	20 +	[2.5,4.0]	50 +
	Nor2a	[0.5,3.2] 或 [8.5,9.2]	25 +	[0.7,2.5]	35 +
	Nor2b	[0.5,10.0]	25 +	[0.4,3.5]	25 +
	Nor2c	[2.0,10.0]	20 +	[0.2,4.0]	20 +
	Nor2d	[2.2,10.0]	20 +	[0.2,4.0]	15 +

表 4.3 对于每个问题挑选参数

	问题 1	问题 2
最大迭代	50	50
群大小	40	40 或 60*
最大速度(V_{max})	3.0	1.0 或 3.0*
样条域广播 r_i	10.0	10.0
统计时间	30	30
适应度函数 f	$f_1 = x_1 + \sin(4x_1) + 1.1x_2\sin(2x_2)$	$f_2 = 100(x_2 - x_1^2)^2 + (x_1 - 1)^2$
范围	[0,10]	[-10,10]
可变最优值	(9.0390,8.6682)	(1,1)
适度最优值	-18.5547	0

注:* 表示仅针对 Nor1,将群大小和最大速度选择为 60 和 3.0。

注意到在 Nor1 中,群体大小和最大速度被选为 60 和 3.0。在剩下的情况中(Uni,Nor2a,Nor2b,Nor2c 和 Nor2d)群体大小和最大速度分别被选为 40 和 1.0。技术 C、D 和 E 中的参数选择和技术 A 相同。

4.9.1.3 仿真

考虑到统计情况,在仿真 30 次后,表 4.4 和表 4.5 分别表示采用五种不同技术和 6 种不同分配的问题 1 和问题 2 的结果。在前 30 次实验中,x_1^*、x_2^* 和f^* 是最合适的第一、第二变量和适应度值;$\bar{x}_1$、$\bar{x}_2$ 和 $\bar{f}$ 是在 30 次仿真实验中的第一、第二变量和适应度值的平均值;σ_{x1}、σ_{x2} 和 σ_f 是在 30 次仿真实验中的第一、第二变量和适应度值的标准差。

表 4.4　问题 1 中每种技术和随机变量的数据

技术		$x_1 \approx 9.0390$			$x_2 \approx 8.6682$			$f \approx -18.5547$		
		x_1^*	$\bar{x}_1$	σ_{x_1}	x_2^*	$\bar{x}_2$	σ_{x_2}	f^*	$\bar{f}$	σ_f
A	Uni	9.0390	8.7775	0.5236	8.6682	7.6290	3.5648	-18.5547	-17.1456	4.2550
	Nor1	9.0505	8.9656	0.0768	8.6427	8.6764	0.0109	-18.5326	-18.1607	0.1523
	Nor2a	9.0758	8.9690	0.0954	8.6896	8.6335	0.0232	-18.4475	-17.8097	0.4676
	Nor2b	9.0336	9.0138	0.0013	8.6267	8.6550	0.0080	-18.5198	-18.2667	0.0560
	Nor2c	9.0408	9.0401	0.0005	8.6726	8.6716	0.0018	-18.5541	-18.4832	0.0048
	Nor2d	9.0364	9.0400	0.0003	8.6711	8.6689	0.0019	-18.5541	-18.4960	0.0048

续表

技术		$x_1 \approx 9.0390$			$x_2 \approx 8.6682$			$f \approx -18.5547$		
		x_1^*	$\bar{x}_1$	σ_{x_1}	x_2^*	$\bar{x}_2$	σ_{x_2}	f^*	$\bar{f}$	σ_f
B	Uni	9.0390	8.6206	0.8377	8.6682	8.4598	0.6288	-18.5547	-17.9063	2.1507
	Nor1	9.0350	8.9373	0.1639	8.6771	8.6576	0.0051	-18.5520	-18.1867	0.2571
	Nor2*a*	9.0295	9.0107	0.0016	8.6392	8.7065	0.0191	-18.5322	-17.0114	0.3496
	Nor2*b*	9.0490	9.0465	0.0022	8.6741	8.6756	0.0064	-18.5467	-18.2815	0.1179
	Nor2*c*	9.0408	9.0401	0.0001	8.6659	8.6716	0.0008	-18.5544	-18.5312	0.0008
	Nor2*d*	9.0410	9.0397	0.0001	8.6689	8.6613	0.0008	-18.5544	-18.5292	0.0005
C	Uni	9.0390	8.8299	0.8030	8.6682	8.2515	1.1677	-18.5547	-17.8859	2.0254
	Nor1	9.0340	9.0326	0.0021	8.6829	8.7025	0.0096	-18.5488	-18.2050	0.1072
	Nor2*a*	9.0629	8.8659	0.3732	8.6395	8.6902	0.0178	-18.4975	-17.8964	0.8067
	Nor2*b*	9.0292	9.0258	0.0019	8.6896	8.6782	0.0048	-18.5390	-18.3223	0.0620
	Nor2*c*	9.0416	9.0499	0.0007	8.6747	8.6671	0.0008	-18.5534	-18.4793	0.0401
	Nor2*d*	9.0399	8.9372	0.3513	8.6699	8.6965	0.0104	-18.5546	-18.0386	1.1760
D	Uni	9.0218	8.6148	4.2615	8.6721	7.9554	1.7334	-18.5331	-15.8799	5.2799
	Nor1	9.0338	8.7265	0.5881	8.6827	8.6631	0.0160	-18.5487	-17.7020	1.0070
	Nor2*a*	9.0364	8.9671	0.0827	8.6644	8.6386	0.0083	-18.5540	-18.2183	0.1292
	Nor2*b*	9.0383	9.0364	0.0002	8.6628	8.6753	0.0010	-18.5541	-18.5184	0.0014
	Nor2*c*	9.0348	9.0381	0.0003	8.6643	8.6711	0.0016	-18.5532	-18.5007	0.0260
	Nor2*d*	9.0422	9.0326	0.0007	8.6739	8.6806	0.0011	-18.5534	-18.4839	0.0170
E	Uni	9.0390	7.8783	3.0638	8.5734	7.4146	3.2577	-18.3432	-15.6750	5.2430
	Nor1	9.0422	8.8912	0.2280	8.6750	8.5826	0.3560	-18.5531	-17.9214	0.6917
	Nor2*a*	9.0340	8.8744	0.2338	8.6697	8.6948	0.0132	-18.5529	-17.9522	0.3418
	Nor2*b*	9.0410	9.0360	0.0004	8.6747	8.6799	0.0018	-18.5536	-18.4921	0.0058
	Nor2*c*	9.0405	9.0419	0.0001	8.6704	8.6604	0.0006	-18.5545	-18.5309	0.0006
	Nor2*d*	9.0387	9.0302	0.0017	8.6664	8.6610	0.0066	-18.5547	-18.3128	0.2683

表 4.5　问题 2 中每种技术和随机变量的数据

技术		$x_1 \approx 9.0390$			$x_2 \approx 8.6682$			$f \approx -18.5547$		
		x_1^*	$\bar{x}_1$	σ_{x_1}	x_2^*	$\bar{x}_2$	σ_{x_2}	f^*	$\bar{f}$	σ_f
A	Uni	1.0000	1.0500	0.0300	1.0000	1.1316	0.2160	0	0.0315	0.0140
	Nor1	1.0000	0.9461	0.0869	1.0000	0.9833	0.2553	0	0.1582	0.0626
	Nor2*a*	1.0000	1.0192	0.0099	1.0000	1.0438	0.0380	0	0.0263	0.0013

续表

技术		$x_1 \approx 9.0390$			$x_2 \approx 8.6682$			$f \approx -18.5547$		
		x_1^*	$\bar{x}_1$	σ_{x_1}	x_2^*	$\bar{x}_2$	σ_{x_2}	f^*	$\bar{f}$	σ_f
A	Nor2b	1.0000	1.0113	0.0030	1.0000	1.0267	0.0129	0	0.0068	0
	Nor2c	1.0000	1.0006	0.0023	1.0000	1.0042	0.0093	0	0.0035	0
	Nor2d	1.0014	1.0005	0.0015	1.0021	1.0032	0.0058	5.1235×10^{-5}	0.0036	0
B	Uni	1.0001	1.0169	0.0163	1.0001	1.0498	0.1040	1.0102×10^{-6}	0.0161	0.0052
	Nor1	1.0415	1.0669	0.0735	1.0849	1.2070	0.4300	1.725×10^{-3}	0.1368	0.0438
	Nor2a	1.0006	0.9890	0.0298	1.0000	1.0059	0.1036	1.4445×10^{-4}	0.0579	0.0063
	Nor2b	1.0000	1.0014	0.0049	1.0000	1.0091	0.0189	0	0.0097	0.0002
	Nor2c	1.0004	1.0060	0.0029	1.0011	1.0137	0.0109	9.1504×10^{-6}	0.0039	0
	Nor2d	1.0040	1.0151	0.0017	1.0080	1.0319	0.0071	1.6026×10^{-5}	0.0030	0
C	Uni	1.0000	1.1024	0.0578	1.0000	1.2708	0.3871	0	0.0693	0.0221
	Nor1	0.9820	0.9399	0.0891	0.9648	0.9728	0.3503	3.4666×10^{-4}	0.1786	0.0403
	Nor2a	1.0000	0.9695	0.0179	1.0000	0.9603	0.0553	0	0.0554	0.0072
	Nor2b	1.0000	1.0055	0.0084	1.0000	1.0221	0.0348	0	0.0175	0.0006
	Nor2c	1.0000	0.9878	0.0045	0.9983	0.9778	0.0172	2.8900×10^{-4}	0.0110	0.0002
	Nor2d	0.9957	1.0286	0.0113	0.9919	1.0689	0.0535	4.1675×10^{-5}	0.0221	0.0013
D	Uni	0.9762	0.7787	0.2269	0.9509	0.8202	0.1250	9.9346×10^{-4}	0.3137	0.7457
	Nor1	1.0000	0.9198	0.0282	1.0000	0.8737	0.0757	0	0.0448	0.0072
	Nor2a	1.0000	0.9515	0.0165	1.0000	0.9216	0.0604	0	0.0306	0.0009
	Nor2b	1.0085	0.9697	0.0053	1.0162	0.9479	0.0186	1.4833×10^{-4}	0.0094	0.0002
	Nor2c	1.0000	0.9749	0.0059	1.0000	0.9551	0.0188	0	0.0077	0.0004
	Nor2d	0.9998	0.9706	0.0058	0.9996	0.9493	0.0217	4.0000×10^{-8}	0.0097	0.0001
E	Uni	0.9827	0.7758	0.2209	0.9670	0.8142	0.2883	4.6847×10^{-4}	0.3126	0.3966
	Nor1	0.9993	0.9908	0.0485	0.9975	1.0281	0.2506	1.2160×10^{-4}	0.0602	0.0130
	Nor2a	1.0139	1.0049	0.0253	1.0293	1.0387	0.0940	3.6398×10^{-4}	0.0471	0.0023
	Nor2b	1.0058	0.9973	0.0031	1.0099	0.9967	0.0122	3.3419×10^{-4}	0.0061	0
	Nor2c	0.9974	0.9927	0.0027	0.9947	0.9885	0.0105	7.8998×10^{-6}	0.0045	0
	Nor2d	1.0020	0.9910	0.0060	1.0023	0.9870	0.0211	2.9436×10^{-4}	0.0103	0.0002

例如,曼德斯[56]、斯克锐和拉蒙[57]在正态分布中使用 0 作为平均值。例如,Nor1 中的工作。根据 4.6.8 节 Nor1、Nor2a,Nor2b,Nor2c 和 Nor2d 分别具有 50%、30.85%、15.87%、2.28% 和 0.13% 的负重量,如图 4.14 所示。

观察表4.4和表4.5中的数据，情况Nor2c（粒子获得2.28%负权重的正态分布下）和Nor2d（粒子获得0.13%负权重的正态分布下）在最佳值和结果的最小方差方面产生一致的结果。将这两种情况与Uni情况进行比较（0~1之间的均匀分布，因此粒子具有严格的正权重），我们得出结论，少量的负速度是有效的，并且在这两种情况下产生了更好的结果。即在该过程中，大多数粒子正在寻找相同的搜索方向（正权重），但是一些粒子正在寻找不同的搜索空间（负权重）。

从两个模拟问题的结果来看，技术A~E得到了令人满意的性能，但看起来技术E成本最高。考虑到相对的可靠性和有效性，我们将研究这些技术在大的问题和CPU时间下的表现。此外，在适当选择一些参数后，使用具有一些负权重的正态分布的粒子确实可以获得更好的结果。因此，我们将研究任何取决于问题的维度的"最佳"粒子群体大小（40），并开发一种自适应算法，其中参数随着算法的进行而自我调整。

4.9.2　多维问题的APSO算法

通过Sphere和Rosenbrock问题这两个例子，证明了所提出的APSO的可靠性[50]。表4.6为针对每个问题选择的参数。对于APSO，如果速度值$V_i^j(t) < V_{\max}$，则$V_i^j(t) = V_{\min}$。$V_{\min}$为最小速度。

表4.6　针对每个问题选择的参数

	Sphere问题	Rosenbrock问题
最大迭代次数	200	200
群大小	40	40
分配	一致	一致
容忍度ε	10^{-8}	10^{-8}
最大速度	0.5	0.5
统计时间	15	15
适应度函数f	$f_1 = \sum_{i=1}^{n} x_i^2$	$f_2 = \sum_{i=1}^{n} [100(x_{i+1} - x_i^2)^2 + (x_i - 1)^2]$
搜索范围$x_i(i=1,\cdots,n)$	$[-10,10]$	$[-10,10]$

图4.24、图4.25和图4.26显示了在Sphere问题上改变和不改变更新速度方向时的APSO和PSO的误差。图4.26描述了在Sphere问题上有+/-标准偏差的APSO和PSO的误差。类似地，图4.27、图4.28和图4.29分别是Rosenbrock问题的模拟。

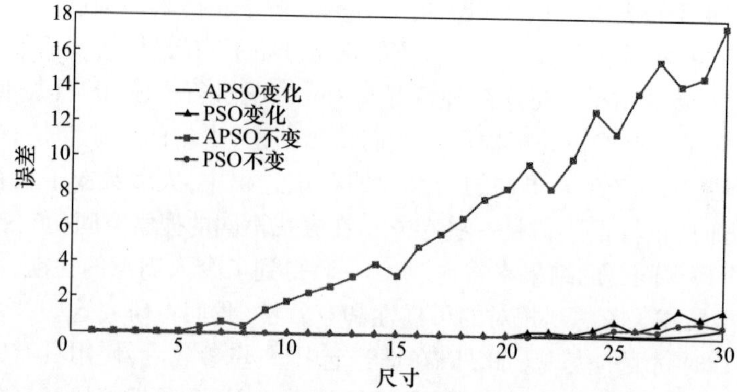

图 4.24　Sphere 问题下的 APSO 和 PSO 误差 Ⅰ

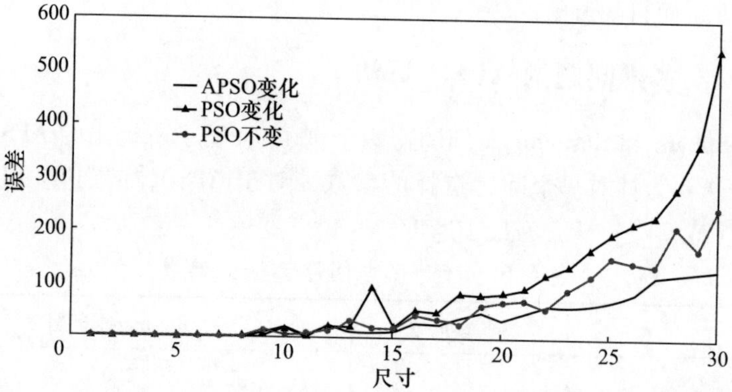

图 4.25　Sphere 问题下的 APSO 和 PSO 误差 Ⅱ

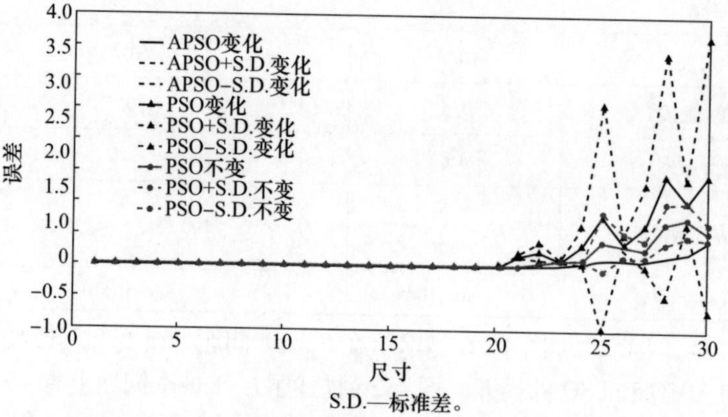

S.D.—标准差。

图 4.26　Sphere 问题标准偏差下的 APSO 和 PSO 误差

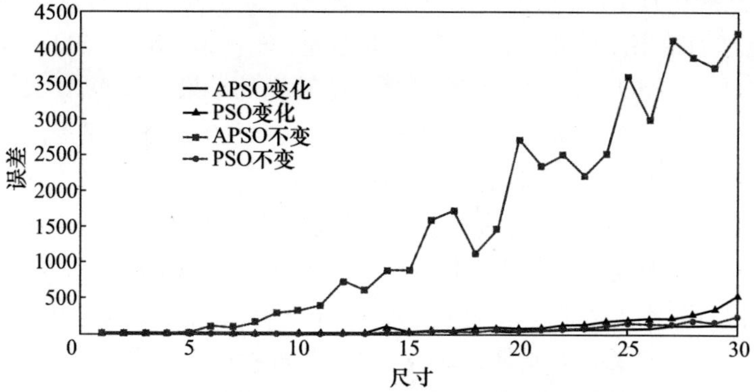

图 4.27　Rosenbrock 问题下的 APSO 和 PSO 误差 Ⅰ

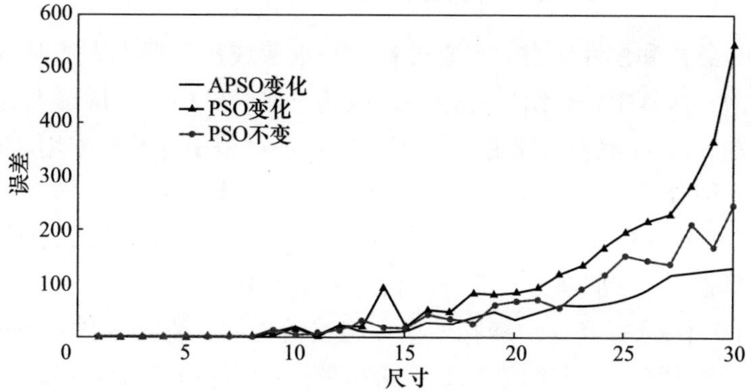

图 4.28　Rosenbrock 问题下的 APSO 和 PSO 误差 Ⅱ

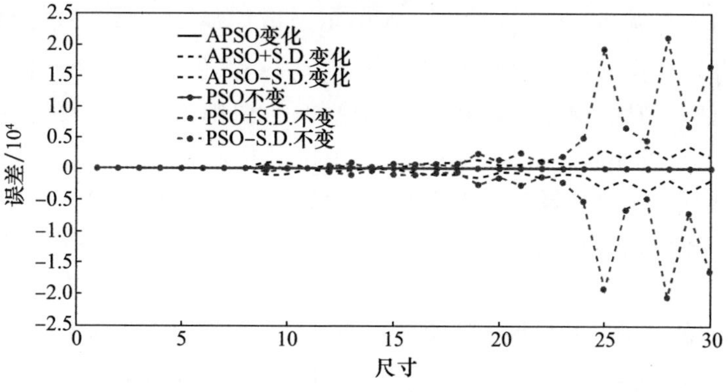

图 4.29　Rosenbrock 问题标准偏差下的 APSO 和 PSO 误差 Ⅱ

基于这些仿真结果，所提出的具有改变更新速度方向的APSO获得了比通用PSO更好的结果。此外，为了比较标准偏差（图4.26和图4.29），具有改变更新速度方向的APSO也产生了比通用PSO更稳定的结果。因为APSO技术试图使用割线平面信息来帮助确定较好的搜索方向，并且对于非常平滑的问题，拥有精确定义的最小值，人们可期望更好的性能。

在探索阶段，最大速度表示粒子位于搜索区域中不感兴趣的区域。改变速度大小及其方向有助于粒子比仅仅改变速度大小更快地逃离这个区域。另一方面，在增强阶段，不太可能出现最大速度，因此不通过使用最大速度分量来改变方向。未来的研究策略可能是，基于群体在其进程中看到的“学习规则”，来恰当地加权自适应割线信息的影响。这将在未来的工作中进行。

4.9.3 PSO在其他生物医学中的应用

炎症是急性和慢性疾病的重要过程。当细菌或病毒侵入人体时，免疫系统就会被激活。这其中一个结果便是白细胞通过血管壁的内皮屏障离开血流，去攻击微生物[58]。细胞表面黏附分子（CAM）在该过程中起基础作用。真核细胞表面上的黏附分子允许在细胞彼此特异性的情况下与细胞外基质相互作用。四个CAM家族介导了大多数黏附相互作用：整合素、钙粘蛋白、免疫球蛋白超家族成员和选择素[59]，它们调节大部分的黏附相互作用。

在CAM中，选择素在白细胞外渗中起关键作用。选择素按三种不同的亚类分类：L－选择蛋白（白细胞）、E－选择蛋白（血管内皮细胞）和P－选择蛋白（血小板或内皮细胞）[60]。L－选择蛋白在大多数白细胞亚群上表达，并负责通过白细胞－白细胞相互作用扩增炎症反应。E－选择蛋白通过转录调节，并响应炎症刺激。P－选择蛋白在细胞内被分割，在激活后早期易位至细胞表面，并且在炎症期间白细胞初始募集至损伤部位中起重要作用。这些选择素不会独立作用，但共同导致炎症。例如，当白细胞附着在血管壁上时，这三种选择素都可能起到将该细胞从血流中拉出的作用。虽然对每种选择蛋白的性质了解很多，但选择蛋白组合对束缚和捕获白细胞的综合反应仍然未知。PSO提供了使用已知的选择素（具有它们各自的配体）的个体破裂特性的机会，并将它们整合到集体情景中，以产生关于炎症调节的创新型的且可测试的假设。不同的动力学和机械性质决定了选择素－白细胞之间的相互作用[60-62]。贝尔模型参数、无应力解离速率和反应顺应性，最初是通过破裂力对加载率对数图的线性区域最小二乘法近似建立的[63]。因此，利用破裂力来捕获大多数分子是重要的。

4.9.3.1 白细胞黏附分子模型

平均破裂力随着加载速率的自然对数线性增加。贝尔[63]首先提出了这种模型。在贝尔的数学模型[62]中，平均破裂力 F_{rup} 表示为

$$F_{rup}=\frac{k_B T}{x_\beta}\ln\left(\frac{x_\beta}{k_{off}^0 k_B T}\right)+\frac{k_B T}{x_\beta}\ln(r_f) \tag{4.9.1}$$

式中：k_B 是玻耳兹曼常数；T 是绝对温度；x_β 表示反应顺应性或机械结合长度；在没有拉力的情况下，k_{off}^0 表示（无应力）解离率；r_f 是加载率。贝尔模型参数 x_β 和 k_{off}^0 描述了 CAM 相互作用的机械特性。因此，单对 CAM 失效的相应概率密度分布通过下式计算：

$$P(F_{rup})=k_{off}^0\exp\left(\frac{x_\beta F_{rup}}{k_B T}\right)\exp\left\{\frac{k_{off}^0 k_B T}{x_\beta r_f}\left[1-\exp\left(\frac{x_\beta F_{rup}}{k_B T}\right)\right]\right\} \tag{4.9.2}$$

基于式(4.9.1)和式(4.9.2)，计算每对 $i(i=1\sim9)$ 的破裂力 $F_{rup,i}$ 和概率 P_i。参数的选择是：玻耳兹曼常数 $k_B=1.38065\times10^{-23}$ J/K；绝对温度 $T=300$K；加载速率 $r_f=2000$pN/s。受体－配体对的所有贝尔参数选择，如表 4.7 所列。因此，总概率 P_{tot} 计算式为

$$P_{tot}(F_{rup}) = \sum_{i=1}^{9} P_i(F_{rup,i}) \tag{4.9.3}$$

表 4.7 每个受体－配体对的贝尔参数

序号	受体－配体对(参考值)	x_β/Å	平均值	k_{off}^0/s^{-1}	平均值
1	L－选择素－PSGL－1[64]	0.16		8.6	
	L－选择素－PSGL－1[60]	1.51	0.835	0.83	4.715
2	L－选择素突变体－PSGL－1[64]	0.15		12.7	
	L－选择素突变体－PSGL－1[64]	0.12		17.3	16.1
	L－选择素突变体－PSGL－1[64]	0.11	0.127	18.3	
3	L－选择素－中性粒细胞[65]	0.24		7.0	
	L－选择素－中性粒细胞[66]	1.11	0.675	2.8	4.9
4	E－选择素－PSGL－1[60]	1.11	1.11	0.24	0.24
5	E－选择素－中性粒细胞[66]	0.18		2.6	
	E－选择素－中性粒细胞[67]	0.31	0.245	0.7	1.65
6	P－选择素－LS174T[61]	0.9	0.9	2.96	2.96
7	P－选择素－PSGL－1[60]	1.35		0.18	
	P－选择素－PSGL－1[64]	0.29		1.1	
	P－选择素－PSGL－1[68]	2.5	1.38	0.022	0.434

续表

序号	受体 - 配体对(参考值)	x_β/Å	平均值	k_{off}^0/s^{-1}	平均值
8	P - 选择素突变体 - PSGL - 1[64]	0.25		1.8	
	P - 选择素突变体 - PSGL - 1[64]	0.33		1.7	
	P - 选择素突变体 - PSGL - 1[64]	0.42	0.33	1.6	1.7
9	P - 选择素 - 中性粒细胞[66]	0.39		2.4	
	P - 选择素 - 中性粒细胞[69]	0.40	0.395	0.93	1.665

图 4.30 显示了每对破裂力下事件计数的概率。PSO 算法的结果显示最佳破裂力 $F_{rup}^*=141.2424$pN,最大概率 $P_{tot}^*=88.4189$。

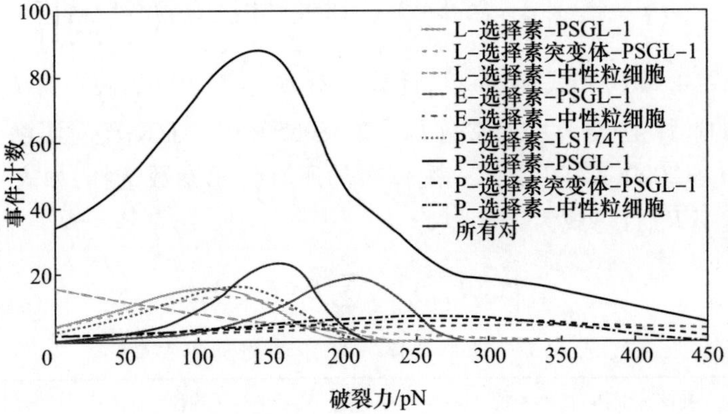

图 4.30　破坏力与对抗计数

这些结果证明了 PSO 可以预测体内产生的多种选择蛋白 - 配体对的综合效应。然后使用这些预测来生成可检验的假设。使用该系统可以帮助理解体内炎症产生过程中,多种选择素 - 配体的表达和调节所起的作用[31,70]。

4.9.4　多维问题的 CHO 算法

通过双曲线和 Rosenbrock 问题两个例子证明所提出的 CHO 的可靠性。TS 的参数选择包括禁忌列表的长度 $N_1=6$,期望列表的长度 $N_2=6$,邻近点的初始半径 $r_1=0.25$,禁忌球的半径 $r_2=0.125$,禁忌球的半径 $r_3=0.06$,最大迭代次数 $I_{max}=200$。GA 的参数选择包括生成的群体大小 = 48,保持交配的染色体数量 = 12,突变率 = 0.04,最大迭代次数 = 200。对于 PSO,参数包括群体大小 = 40,最大速度 = 2.5,均匀分布被选为随机变量,容忍度 $\varepsilon=10^{-8}$,最大迭代次数 = 200。建议的 CHO 利用 TS 和 PSO 的相同参数选择和期望区域的恒定半径 $r_b=0.5$[23]。

双曲线和 Rosenbrock 问题的成本函数分别描述为

$$f_{\text{Hyperbolic}} = \sum_{i=1}^{n} x_i^2 \tag{4.9.4}$$

$$f_{\text{Rosenbrock}} = \sum_{i=1}^{n} [100\,(x_{i+1} - x_i^2)^2 + (x_i - 1)^2] \tag{4.9.5}$$

表 4.8 和表 4.9 分别显示了在 30 次模拟后，五种算法（ECTS、CGA、PSO、ECTS + CGA 和 CHO）对于双曲线和 Rosenbrock 问题的平均值和标准差误差。两个问题的搜索范围都是[-10,10]。根据仿真结果，PSO 可以在小于 10 维的双曲线问题上获得优异的性能。

表 4.8　双曲问题的五种算法的均值和标准差误差

n	ECTS	CGA	PSO	ECTS + CGA	CHO
2	0.0179 ±0.0218	0.0002 ±0.0007	$3.1\times10^{-8}\pm3.8\times10^{-8}$	$0.0001\pm3.9\times10^{-5}$	$1.5\times10^{-9}\pm1.5\times10^{-9}$
3	0.0190 ±0.0146	0.0028 ±0.0041	$1.2\times10^{-8}\pm8.0\times10^{-7}$	0.0026 ±0.0061	$1.1\times10^{-7}\pm2.1\times10^{-7}$
5	0.0238 ±0.0261	0.0471 ±0.0265	0.0001 ±0.0001	0.0207 ±0.0260	$4.7\times10^{-6}\pm1.0\times10^{-5}$
10	0.0303 ±0.0327	0.6225 ±0.3012	0.0011 ±0.0014	0.0342 ±0.0285	0.0003 ±0.0003
20	0.0501 ±0.0506	3.8993 ±0.8040	0.2263 ±0.1611	0.0432 ±0.0400	0.0297 ±0.0903
30	0.0649 ±0.0782	8.1319 ±1.1536	2.0799 ±1.0858	0.1035 ±0.0760	0.2698 ±0.6996
50	0.1339 ±0.1240	16.1665 ±1.4039	10.8841 ±2.2802	0.1273 ±0.1276	0.4632 ±2.0628
100	0.2383 ±0.2189	38.0997 ±1.8511	39.8710 ±3.9074	0.2073 ±0.2442	1.2373 ±5.9897
300	0.6314 ±0.6334	129.7538 ±3.0646	176.5882 ±9.2199	0.8907 ±0.9396	0.4526 ±0.3736
500	1.1307 ±1.0052	222.8210 ±5.9827	318.8833 ±11.5058	1.0688 ±0.9051	1.0291 ±0.9075

表 4.9　Rosenbrock 问题的五种算法的均值和标准差误差

n	ECTS	CGA	PSO	ECTS + CGA	CHO
2	0.0137 ±0.0138	0.5242 ±0.2789	0.0003 ±0.0004	0.2703 ±0.7590	0.0336 ±0.1327
3	0.0148 ±0.0151	1.0620 ±0.3752	0.6215 ±1.0571	0.1653 ±0.0796	0.0891 ±0.1511
5	0.0195 ±0.0203	2.1632 ±0.9059	2.2764 ±0.9663	0.1326 ±0.0486	0.1561 ±0.2368
10	0.0373 ±0.0397	4.2154 ±2.5281	7.9283 ±1.8828	0.0619 ±0.0279	0.5464 ±1.8083
20	0.0568 ±0.0568	8.0005 ±3.1585	16.4347 ±3.4651	0.0535 ±0.0209	0.0893 ±0.0486
30	0.0649 ±0.0578	12.4630 ±3.9880	23.0807 ±3.5200	0.0720 ±0.0315	2.4114 ±6.9677
50	0.1233 ±0.1500	19.8758 ±4.3163	39.4033 ±4.3694	0.0984 ±0.0436	1.7678 ±8.7988
100	0.2333 ±0.1938	43.0156 ±4.9253	88.2320 ±6.8456	0.1769 ±0.1233	0.2243 ±0.0775
300	0.4577 ±0.4289	145.6715 ±7.2527	304.8118 ±15.5815	0.5147 ±0.4542	3.6957 ±10.7818
500	1.4962 ±1.8032	246.2241 ±8.9461	531.4223 ±14.6164	0.8567 ±0.8093	1.0540 ±0.8991

然而，PSO 无法在大于 SO 维的双曲线问题上找到全局最小值。同样，GA

可以在低于 IO 维的双曲线问题上获得良好的性能，但 GA 的表现不如 PSO。此外，GA 不能达到大于 30 维双曲线问题的全局最小值。

对于所有维度双曲线问题，TS 有较好的鲁棒性，但是在小于 10 维的双曲线问题上，PSO 比 TS 好得多。两种混合算法，ECTS + CGA 和 CHO，都将 TS 和 GA 或 PSO 的优势结合在整个双曲线问题上。此外，在小于 20 维的双曲线问题上，CHO 比 ECTS + GA 更稳健。

对于 Rosenbrock 问题，仿真结果表明，得到了与双曲线问题类似的结果。然而，PSO 和 GA 在寻找全局最小值方面更加困难，特别是在更高维 Rosenbrock 问题上。TS 仍然保留了强大的特性。类似地，除了 300 维 Rosenbrock 问题之外，混合算法 ECTS + CGA 和 CHO 都将 TS 和 GA 或 PSO 的优势结合在所有维 Rosenbrock 问题上。在更高维的 Rosenbrock 问题上，CHO 显示出对从有希望的列表中选择最有希望的区域非常敏感的依赖性。

这项工作显示了五种算法（ECTS、CGA、PSO、ECTS + CGA 和 CHO）对多维双曲线和 Rosenbrock 问题的比较。仿真结果表明，所提出的 CHO 算法结合了 TS 和 PSO 的优点，获得了稳健的结果。然而，在更高维问题上，CHO 显示出对从有希望的列表中选择最有希望的区域的敏感依赖性。选择有希望区域的恒定半径（r_b）起着关键作用。因此，在将来的工作中研究该参数选择。

参考文献

[1] F. O. Karray and C. De Silva. *Soft Computing and Intelligent Systems Design: Theory, Tools and Applications.* Pearson Educational Limited, Harlow, UK, 2004.

[2] A. Konar. *Computational Intelligence: Principles, Techniques and Applications.* Springer – Verlag, Berlin, Germany, 2005.

[3] R. R. Seeley, T. D. Stephens, and P. Tate. *Anatomy and Physiology, Eighth Edition.* The McGraw – Hill, New York, USA, 2007.

[4] J. – S. R. Jang, C. – T. Sun, and E. Mizutani. *Neuro – Fuzzy and Soft Computing: A Computational Approach to Learning and Machine Intelligence.* Prentice Hall PTR, Upper Saddle River, New Jersey, USA, 1997.

[5] H. T. Nguyen, N. R. Prasad, C. L. Walker, and E. A. Walker. *A First Course in Fuzzy and Neural Control.* Chapman & Hall/CRC, Boca Raton, Florida, USA, 2003.

[6] F. L. Lewis, D. M. Dawson, and C. T. Abdallah. *Robot Manipulators Control: Second Edition, Revised and Expanded.* Marcel Dekker, Inc., New York, USA, 2004.

[7] L. A. Zadeh. Fuzzy sets. *Information and Control*, 8: 338 – 353, 1965.

[8] S. Sumathi and Surekha Paneerselvam. *Computational Intelligence Paradigms: Theory and Applications Using Matlab.* CRC Press, Boca Raton, Florida, USA, 2010.

[9] E. H. Mamdani and S. Assilian. An experiment in linguistic synthesis with a fuzzy logic controller. *International Journal of Man – Machine Studies*, 7(1): 1 – 13, 1975.

[10] T. Takagi and M. Sugeno. Fuzzy identification of systems and its application tomodeling and control. *IEEE Transactions on Systems, Man, and Cybernetics*, 15(1):116 - 132, 1985.

[11] F. L. Lewis, S. Jagannathan, and A. Yesildirek. *Neural Network Control of Robotic Manipulators and Nonlinear Systems.* Taylor & Francis, London, UK, 1999.

[12] B. Widrow and M. E. Hoff. Adaptive switching circuits. *IRE WESCON Convention Record*, pp. 96 - 104, 1960.

[13] H. Takagi. Fusion technology of neural networks and fuzzy systems: A chronicled progression from the laboratory to our daily lives. *International Journal of Applied Mathematics and Computer Science*, 10(4):647 - 673, 2000.

[14] J. - S. R. Jang. ANFIS: Adaptive - network - based fuzzy inference system. *IEEE Transactions on Systems, Man, and Cybernetics*, 23(665 - 685):3 - 12, 1993.

[15] C. - H. Chen, D. S. Naidu, and M. P. Schoen. Adaptive control for a five - fingered hand with unknown mass and inertia. *World Scientific and Engineering Academy and Society (W SEAS) Journal on Systems*, 10(5): 148 - 161, May 2011.

[16] C. - H. Chen and D. S. Naidu. "Fusion of fuzzy logic and PD control for a fivefingered smart prosthetic hand," in *Proceedings of the 2011 IEEE International Conference on Fuzzy Systems (FUZZ - IEEE 2011)*, pp. 2108 - 2115, Taipei, Taiwan, June 27 - 30, 2011.

[17] C. - H. Chen and D. S. Naidu. Hybrid control strategies for a five - finger robotic hand. *Biomedical Signal Processing and Control*, 8(4):382 - 390, July 2013.

[18] F. Glover. Tabu search - part I. *ORSA Journal of Computing*, 10(3):190 - 205, 1989.

[19] F. Glover. Tabu search - part IL *ORSA Journal of Computing*, 2(1):4 - 32, 1990.

[20] P. Siarry and G. Berthiau. Fitting tabu search to optimize functions of continuous variables. *International Journal for Numerical Methods in Engineering*, 40:2449 - 2459, 1997.

[21] R. Chelouah and P. Siarry. Tabu search applied to global optimization. *European Journal of Operational Research*, 123:256 - 270, 2000.

[22] C. - H. Chen, M. P. Schoen, and K. W. Bosworth. "A condensed hybrid optimization algorithm using enhanced continuous tabu search and particle swarm optimization," in *Proceedings of the ASME 2009 Dynamic Systems and Control Conference (DSCC)*, Hollywood, California, USA, October 12 - 14, 2009 (No. DSCC2009 - 2526).

[23] R. Chelouah and P. Siarry. A continuous genetic algorithm design for global optimization of multimodal functions. *Journal of Heuristics*, 6:191 - 213, 2000.

[24] C. - H. Chen and D. S. Naidu. "Hybrid genetic algorithm PID control for a five - fingered smart prosthetic hand," in *Proceedings of the Sixth International Conference on Circuits, Systems and Signals (CSS' 11)*, pp. 57 - 63, Vouliagmeni Beach, Athens, Greece, March 7 - 9, 2012.

[25] J. Kennedy and R. Eberhart. "Particle swarm optimization," in *IEEE lnternational Conference on Neural Networks*, 4:1942 - 1948, 1995.

[26] J. Kennedy, R. Eberhart, and Y. Shi. *Swarm intelligence.* Morgan Kaufmann Publishers, San Francisco, California, USA, 2001.

[27] A. P. Engelbrecht. *Fundamentals of Computational Swarm Intelligence.* John Wiley & Sons, 2006.

[28] M. Clerc. *Particle Swarm Optimization.* ISTE Publishing Company, 2006.

[29] R. C. Eberhart and Y. Shi. *Computational Intelligence Concepts to Implementations.* Morgan Kaufmann Publishers,2007.

[30] A. P. Engelbrecht. *Computationa Intelligence:An Introduction,Second Edition.* John Wiley & Sons,2007.

[31] C. – H. Chen, K. W. Bosworth, M. P. Schoen, S. E. Bearden, D. S. Naidu, and Perez – Gracia. "A study of particle swarm optimization on leukocyteadhesion molecules and control strategies for smart prosthetic hand," in *2008 IEEE Swarm Intelligence Symposium* (*IEEE SIS08*), St. Louis, Missouri, USA, September 21 – 23,2008.

[32] C. – H. Chen, D. S. Naidu, A. Perez – Gracia, and M. P. Schoen. "Fusion of hard and soft control techniques for prosthetic hand," in *Proceedings of the International Association of Science and Technology for Development* (*IASTED*) *International Conference on Intelligent Systems and Control* (*ISC 2008*), pp. [a]125, Orlando, Florida, USA, November 16 – 18,2008.

[33] Y. Shi and R. Eberhart. "Parameter selection in particle swarm optimization," in *Evolutionary Programming VII:Proceedings of the Seventh Annual Conference on Evolutionary Programming*, pp. 591 – 600, New York, USA,1998.

[34] Y. Shi and R. Eberhart. "Empirical study of particle swarm optimization," in *Proceedings of the IEEE Congress on Evolutionary Computation* (*IEEE Press*), pp. 1945 – 1950,1999.

[35] R. Eberhart and Y. Shi. "Comparing inertia weights and constriction factors in particle swarm optimization," in *Proceedings of the IEEE Congress on Evolutionary Computation*, pp. 84 – 88, San Diego, California, USA, 2000.

[36] M. Clerc and J. Kennedy. The particle swarm – explosion, stability, and convergence in a multi – dimensional complex space. *IEEE Transactions on Evolutionary Computation*,6(1):58 – 73,2002.

[37] H. J. Meng, P. Zheng, R. Y. Wu, X. J. Hao, and Z. Xie. "A hybrid particle swarm algorithm with embedded chaotic search," in *Proceedings of the IEEE Conference on Cybernetics and Intelligent Systems*, pp. 367 – 371, Singapore, 2004.

[38] S. H. Ho, S. Yang, G. Ni, E. W. C. Lo, and H. C. Wong. A particle swarm optimization – based method for multiobjective design optimizations. *IEEE Transactions on Magnetics*,41(5):1756 – 1759,2005.

[39] J. H. Seo, C. H. Im, C. G. Heo, J. K. Kim, H. K. Jung, and C. G. Lee. Multimodal function optimization based on particle swarm optimization. *IEEE Transactions on Magnetics*,42(4):1095 – 1098,2006.

[40] R. A. Krohling and L. dos S. Coelho. Coevolutionary particle swarm optirnization using gaussian distribution for solving constrained optimization problems. *IEEE Transactions on Systems, Man, and Cybernetics, Part B: Cybernetics*,36(6):1407 – 1416,2006.

[41] R. Brits, A. P. Engelbrecht, and F. van den Bergh. Locating multiple optima using particle swarm optimization. *Applied Mathematics and Computation*,189(2):1859 – 1883,2007.

[42] F. van den Bergh and A. P. Engelbrecht. A study of particle swarm optimization particle trajectories. *Information Sciences*,176(8):937 – 971,2006.

[43] H. Someya. "Cautious particle swarm," in *2008 IEEE Swarm Intelligence Symposium*, St. Louis, Missouri, USA, September 21 – 23,2008.

[44] L. – Y. Chuang, S. – W. Tsai, and C. – H. Yang. "Catfish particle swarm optirnization," in *2008 IEEE Swarm Intelligence Symposium*, St. Louis, Missouri, USA, September 21 – 23,2008.

[45] F. Pan, X. Hu, R. Eberhart, and Y. Chen. "An analysis of bare bones particle swarm," in *2008 IEEE Swarm*

Intelligence Symposium, St. Louis, Missouri, USA, September 21 – 23, 2008.

[46] C. – H. Chen, K. W. Bosworth, and M. P. Schoen. "Investigation of particleswarm optimization dynamics," in *Proceedings of International Mechanical Engineering Congress and Exposition (IMECE) 2007*, Seattle, Washington, USA, November 11 – 15, 2007 (No. IMECE2007 – 41343).

[47] R. Eberhart and Y. Shi. Guest editorial special on particle swarm optimization. *IEEE Transactions on Evolutionary Computation*, 8(3): 201 – 203, 2004.

[48] S. N. Atluri, H. G. Kim, and J. Y. Cho. A critical assessment of the truly Meshless Local Petrov – Galerkin (MLPG), and Local Boundary Integral Equation (LBIE) methods. *Computational Mechanics*, 24: 348 – 372, 1999.

[49] W. Pan, F. Zhun, F. Shan, and Z. Yun. "Study on a novel hybrid adaptive genetic algorithm embedding conjugate gradient algorithm," in *Proceedings of the Third World Congress on Intelligent Control and Automation*, Hefei, P. R. China, June 28 – July 2, 2000.

[50] C. – H. Chen, K. W. Bosworth, and M. P. Schoen. "An adaptive particle swarm method to multiple dimensional problems," in *Proceedings of the International Association of Science and Technology for Development (IASTED) International Symposium on Computational Biology and Bio informatics (CBB 2008)*, pp. 260 – 265, Orlando, Florida, USA, November 16 – 18, 2008.

[51] D. E. Goldberg. *Genetic Algorithm in Search, Optimization and Machine Learning*. Addison Wesley Longman, Inc., 1989.

[52] Y. Guo, L. – P. Chen, S. Wang, and J. Zhao. A new simulation optirmzation system for the parameters of a machine cell simulation model. *International Journal of Advanced Manufacturing Technology*, 21: 620 – 626, 2003.

[53] L. Pladugu, M. P. Schoen, and B. Williams. "Intelligent techniques for starpattern recognition," in *Proceedings of IMECE 2003*, Washington, District of Columbia, USA, November 16 – 17, 2003.

[54] B. Rammar, M. P. Schoen, F. Lin, and B. G. Williams. "Hybrid optimization algorithm using enhanced continuous tabu search and genetic algorithm for parameter estimation," in *Proceedings of International Mechanical Engineering Congress and Exposition (IMECE) 2006*, Chicago, Illinois, USA, November5 – 10, 2006 (No. IMECE2006 – 13374).

[55] B. Ramkumar, M. P. Schoen, and F. Lin. "App! ication of an intelligent hybrid optimization technique for parameter estl matl on in the presence of colored noise," in *Proceedings of International Mechanical Engineering Congress and Exposition (IMECE) 2007*, Seattle, Washington, USA, November 11 – 15, 2007 (No. IMECE2007 – 41352).

[56] R. Mendes, J. Kennedy, and J. Neves. "Watch thy neighbor or how the swarm can learn from its environment," in *Proceedings of the IEEE Swarm Intelligence Symposium 2003 (SIS 2003)*, pp. 88 – 94, Indianapolis, Indiana, USA, 2003.

[57] B. R. Secrest and G. B. Lamont. "Visualizing particle swarm optimizationxgaussian particle swarm optimization," in *Proceedings of the IEEE Swarm Intelligence Symposium 2003 (SIS 2003)*, pp. 198 – 204, Indianapolis, Indiana, USA, 2003.

[58] P. J. Delves and I. M. Roitt. Advances in immunology: The immune system. *The New England Journal of Medicine*, 343(1): 37 – 50, July 6, 2000.

[59] R. Rhoades and R. Pflanzer. *Human Physiology*. Thomson Brooks/Cole, Fourth Edition, 2003.

[60] W. D. Hanley, D. Wirtz, and K. Konstantopoulos. Distinct kinetic and mechanical properties govern selectin – leukocyte interactions. *Journal of Cell Science*, 117(12):2503 – 2511, 2004.

[61] W. Hanley, O. McCarty, S. Jadhav, Y. Tseng, D. Wirtz, and K. Konstantopoulos. Single molecule characterization of p – selectin/ligand binding. *The Journal of Biological Chemistry*, 278(12):10556 – 10561, 2003.

[62] J. te Riet, A. W. Zimmerman, A. Cambi, B. Joosten, S. Speller, R. Torensma, F. N. van Leeuwen, C. G. Figdor, and F. de Lange. Distinct kinetic and mechanical properties govern alcam – mediated interactions as shown by singlemolecule force spectroscopy. *Journal of Cell Science*, 120(22):3965 – 3976, 2007.

[63] G. I. Bell. Models for the specific adhesion of cells to cells. *Science*, 200:618 – 627, 1978.

[64] V. Ramachandran, M. U. Nollert, W. – J. Liu, H. Qiu, R. D. Cummings, C. C. Zhu, and R. P. McEver. Tyrosine replacement in p – selectin glycoprotein ligand – I affects distinct kinetic and mechanical properties of bonds with pand 1 – selectin. *Proceedings of the National Academy of Sciences of the United States of America*, 96(24):13771 – 13776, 1999.

[65] R. Alon, S. Chen, R. Fuhlbrgge, K. D. Puri, and T. A. Springer. The kinetics and shear threshold of transient and rolling interactions of 1 – selectin with its ligand on leukocytes. *Proceedings of the National Academy of Sciences of the United States of America*, 95(20):11631 – 11636, 1998.

[66] M. J. Smith, E. L. Berg, and M. B. Lawrence. A direct comparison of selectinmediated transient, adhesive events using high temporal resolution. *Biophysical Journal*, 77(6):3371 – 3383, 1999.

[67] R. Alon, S. Chen, K. D. Puri, E. B. Finger, and T. A. Springer. The kinetics of L – selectin tethers and the mechanics of selectin – mediated rolling. *The Journal if Cell Biology*, 138(5):1169 – 1180, 1997.

[68] J. Fritz, A. G. Katopodis, F. Kolbinger, and D. Anselmetti. Force – mediatedkinetics of single P – selectin/ligand complexes observed by atomic force microscopy. *Proceedings of the National Academy of Sciences of the United States America*, 95(21):12283 – 12288, 1998.

[69] R. Alon, D. A. Hammer, and T. A. Springer. Lifetime of the P – selectincarbohydrate bond and its response to tensile force in hydrodynamic flow. *Nature (London)*, 374(6522):539 – 542, 1995.

[70] C. – H. Chen. *Hybrid Control Strategies for Smart Prosthetic Hand.* PhD Dissertation, Idaho State University, Pocatello, Idaho, USA, May 2009.

第 5 章　硬控制策略和软控制策略的融合 I

本章是第 4 章的延续，重点介绍硬控制策略和软控制策略在机械手/假肢上的融合（混合或集成）应用。尤其是 PID 控制、最优控制和自适应硬控制技术与 ANFIS 软控制技术在轨迹规划上的集成。5.1 节对反馈线性化进行简要介绍，5.2 节中介绍 PID 控制，5.3 节介绍最优控制，5.4 节介绍自适应控制。模拟结果在 5.5 节中给出。

5.1　反馈线性化

假设式(3.3.16)中的非线性项 $\boldsymbol{N}(\boldsymbol{q},\dot{\boldsymbol{q}})$ 仅模拟科里奥利力/向心力和重力项，那么拇指和所有手指的动态方程重写如下：

$$\boldsymbol{M}(\boldsymbol{q}(t)\ddot{\boldsymbol{q}}(t))+\boldsymbol{N}(\boldsymbol{q}(t),\dot{\boldsymbol{q}}(t))=\boldsymbol{\tau}(t) \tag{5.1.1}$$

式中：$\boldsymbol{N}(\boldsymbol{q}(t),\dot{\boldsymbol{q}}(t))=\boldsymbol{C}(\boldsymbol{q}(t),\dot{\boldsymbol{q}}(t))+\boldsymbol{G}(\boldsymbol{q}(t))$ 表示非线性项。

通过使用反馈线性化技术[1-2]找到一种变换方式，将式(5.1.1)表示的非线性动力学系统转换为线性状态变量系统。将关节的角位置/速度状态 $\boldsymbol{x}(t)$ 定义为下式可得到动态的替代状态空间方程[2-3]，即

$$\boldsymbol{x}(t)=[\boldsymbol{q}'(t)\ \dot{\boldsymbol{q}}'(t)]' \tag{5.1.2}$$

动态模型不变并将式(5.1.1)写成如下形式：

$$\frac{\mathrm{d}}{\mathrm{d}t}\dot{\boldsymbol{q}}(t)=\ddot{\boldsymbol{q}}(t)=-[\boldsymbol{M}(\boldsymbol{q}(t))]^{-1}\boldsymbol{N}(\boldsymbol{q}(t),\dot{\boldsymbol{q}}(t))+[\boldsymbol{M}(\boldsymbol{q}(t))]^{-1}\boldsymbol{\tau}(t) \tag{5.1.3}$$

5.1.1　状态变量表示方法

形式 1：选择式(5.1.2)所示的状态变量，通常情况下，如式(5.1.3)表示的对于一根机械手指的机械手的动力学方程在状态空间形式上可写成如下形式

$$\dot{\boldsymbol{x}}(t)=\begin{bmatrix}\dot{\boldsymbol{q}}(t)\\ \ddot{\boldsymbol{q}}(t)\end{bmatrix}$$

$$=\begin{bmatrix}\dot{\boldsymbol{q}}(t)\\ -[\boldsymbol{M}(\boldsymbol{q}(t))]^{-1}\boldsymbol{N}(\boldsymbol{q}(t),\dot{\boldsymbol{q}}(t))\end{bmatrix}+\begin{bmatrix}\boldsymbol{0}\\ [\boldsymbol{M}(\boldsymbol{q}(t))]^{-1}\end{bmatrix}\boldsymbol{\tau}(t) \tag{5.1.4}$$

这是一种典型的非线性系统形式

$$\dot{\boldsymbol{x}}(t)=\boldsymbol{f}(\boldsymbol{x}(t),\boldsymbol{u}(t),t) \tag{5.1.5}$$

式中：$\boldsymbol{u}(t)=\boldsymbol{\tau}(t)$。

形式 2：可以用另一种通用的状态空间形式写式(3.3.16)，即

$$\begin{bmatrix}\boldsymbol{I} & \boldsymbol{0}\\ \boldsymbol{0} & \boldsymbol{M}(\boldsymbol{q}(t))\end{bmatrix}\begin{bmatrix}\dot{\boldsymbol{q}}(t)\\ \ddot{\boldsymbol{q}}(t)\end{bmatrix}=\begin{bmatrix}\dot{\boldsymbol{q}}(t)\\ -\boldsymbol{N}(\boldsymbol{q}(t),\dot{\boldsymbol{q}}(t))\end{bmatrix}+\begin{bmatrix}\boldsymbol{0}\\ \boldsymbol{I}\end{bmatrix}\boldsymbol{\tau}(t) \tag{5.1.6}$$

另一方面，通过式(5.1.2)和式(5.1.3)的描述，式(3.3.16)的布鲁诺夫斯基规范形式的另一个替代线性状态变量方程被写为

$$\dot{\boldsymbol{x}}(t)=\begin{bmatrix}\boldsymbol{0} & \boldsymbol{I}\\ \boldsymbol{0} & \boldsymbol{0}\end{bmatrix}\boldsymbol{x}(t)+\begin{bmatrix}\boldsymbol{0}\\ \boldsymbol{I}\end{bmatrix}\boldsymbol{u}(t) \tag{5.1.7}$$

由下式得到其控制输入矢量，即

$$\boldsymbol{u}(t)=-[\boldsymbol{M}(\boldsymbol{q}(t))]^{-1}\boldsymbol{N}(\boldsymbol{q}(t),\dot{\boldsymbol{q}}(t))+[\boldsymbol{M}(\boldsymbol{q}(t))]^{-1}\boldsymbol{\tau}(t) \tag{5.1.8}$$

假设机械手需要跟踪在路径生成或跟踪时的期望轨迹 $\boldsymbol{q}_{\mathrm{d}}(t)$。那么，跟踪误差 $\boldsymbol{e}(t)$ 定义为

$$\boldsymbol{e}(t)=\boldsymbol{q}_{\mathrm{d}}(t)-\boldsymbol{q}(t) \tag{5.1.9}$$

式中：$\boldsymbol{q}_{\mathrm{d}}(t)$ 是关节的理想角度矢量，由式(2.5.2)、式(2.3.4)和式(2.3.7)得到；$\boldsymbol{q}(t)$ 是关节的实际角度矢量。对式(5.1.9)求两次微分，得到

$$\dot{\boldsymbol{e}}(t)=\dot{\boldsymbol{q}}_{\mathrm{d}}(t)-\dot{\boldsymbol{q}}(t),\quad \ddot{\boldsymbol{e}}(t)=\ddot{\boldsymbol{q}}_{\mathrm{d}}(t)-\ddot{\boldsymbol{q}}(t) \tag{5.1.10}$$

将式(5.1.2)代入式(5.1.10)得到

$$\ddot{\boldsymbol{e}}(t)=\ddot{\boldsymbol{q}}_{\mathrm{d}}(t)+[\boldsymbol{M}(\boldsymbol{q}(t))]^{-1}[\boldsymbol{N}(\boldsymbol{q}(t),\dot{\boldsymbol{q}}(t))-\boldsymbol{\tau}(t)] \tag{5.1.11}$$

控制功能定义为

$$\boldsymbol{u}(t)=\ddot{\boldsymbol{q}}(t)+[\boldsymbol{M}(\boldsymbol{q}(t))]^{-1}[\boldsymbol{N}(\boldsymbol{q}(t),\dot{\boldsymbol{q}}(t))-\boldsymbol{\tau}(t)] \tag{5.1.12}$$

这通常被称为反馈线性化控制律，它也经常转换成如下形式：

$$\boldsymbol{\tau}(t)=\boldsymbol{M}(\boldsymbol{q}(t))[\ddot{\boldsymbol{q}}_{\mathrm{d}}(t)-\boldsymbol{u}(t)]+\boldsymbol{N}(\boldsymbol{q}(t),\dot{\boldsymbol{q}}(t)) \tag{5.1.13}$$

利用式(5.1.10)和式(5.1.12)之间的联系,并定义状态矢量,跟踪误差动力学方程为

$$\dot{\boldsymbol{x}}(t)=\begin{bmatrix}\mathbf{0} & \boldsymbol{I}\\ \mathbf{0} & \mathbf{0}\end{bmatrix}\boldsymbol{x}(t)+\begin{bmatrix}\mathbf{0}\\ \boldsymbol{I}\end{bmatrix}\boldsymbol{u}(t) \tag{5.1.14}$$

请注意,这是一个形如下式的线性系统,即

$$\dot{\boldsymbol{x}}(t)=\boldsymbol{A}\boldsymbol{x}(t)+\boldsymbol{B}\boldsymbol{u}(t) \tag{5.1.15}$$

其控制输入矢量由下式得到:

$$\boldsymbol{u}(t)=-\boldsymbol{M}^{-1}(\boldsymbol{q}(t))[\boldsymbol{N}(\boldsymbol{q}(t),\dot{\boldsymbol{q}}(t))-\boldsymbol{\tau}(t)] \tag{5.1.16}$$

然后通过下式计算,得到所有关节的所需扭矩,即

$$\boldsymbol{\tau}(t)=\boldsymbol{M}(\boldsymbol{q}(t))\boldsymbol{u}(t)+\boldsymbol{N}(\boldsymbol{q}(t),\dot{\boldsymbol{q}}(t)) \tag{5.1.17}$$

5.2 PD/PI/PID 控制

比例(PD)控制器是最简单的闭环控制器,用于控制机器人操纵器。PD 控制器利用的是比例部分(位置)和微分部分(速度)的反馈组合。但是,如果机器人操纵器动力学包含重力项的特定矢量(如式(3.3.15)和式(3.4.5)),那么简单的 PD 控制律[4]就无法实现位置控制目标。因此,为了满足位置控制目标,在比例积分(PI)和比例积分微分(PID)控制器中加入一种积分分量。PD、PI 和 PID 控制器在本节[5]中作简要描述。

5.2.1 PD 控制器

图 5.1 显示了 PD 控制器的框图[2]。结合比例和导数对角线增益矩阵,控制信号 $\boldsymbol{u}(t)$ 变为

$$\boldsymbol{u}(t)=-\boldsymbol{K}_{\mathrm{p}}\boldsymbol{e}(t)-\boldsymbol{K}_{\mathrm{D}}\dot{\boldsymbol{e}}(t) \tag{5.2.1}$$

紧接着定义闭环误差动态表达式和状态空间表达式,如下:

$$\ddot{\boldsymbol{e}}(t)+\boldsymbol{K}_{\mathrm{D}}\dot{\boldsymbol{e}}(t)+\boldsymbol{K}_{\mathrm{p}}\boldsymbol{e}(t)=0 \tag{5.2.2}$$

$$\frac{\mathrm{d}}{\mathrm{d}t}\begin{bmatrix}\boldsymbol{e}(t)\\ \dot{\boldsymbol{e}}(t)\end{bmatrix}=\begin{bmatrix}\mathbf{0} & \boldsymbol{I}\\ -\boldsymbol{K}_{\mathrm{p}} & -\boldsymbol{K}_{\mathrm{D}}\end{bmatrix}\begin{bmatrix}\boldsymbol{e}(t)\\ \dot{\boldsymbol{e}}(t)\end{bmatrix}+\begin{bmatrix}\mathbf{0}\\ \boldsymbol{I}\end{bmatrix}\boldsymbol{u}(t) \tag{5.2.3}$$

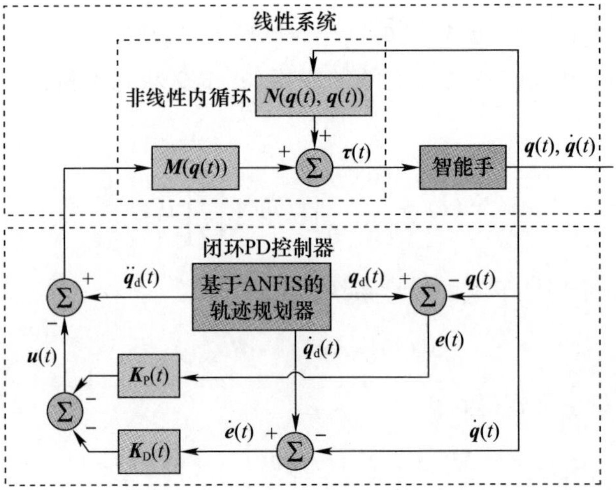

图 5.1　机器人手的融合 PD 控制器框图

那么

$$\boldsymbol{\tau}(t)=\boldsymbol{M}(\boldsymbol{q}(t))[\ddot{\boldsymbol{q}}(t)-\boldsymbol{u}(t)]+\boldsymbol{N}(\boldsymbol{q}(t),\dot{\boldsymbol{q}}(t)) \tag{5.2.4}$$

或者为

$$\boldsymbol{\tau}(t)=\boldsymbol{M}(\boldsymbol{q}(t))[\ddot{\boldsymbol{q}}(t)+\boldsymbol{K}_{\mathrm{D}}\dot{\boldsymbol{e}}(t)+\boldsymbol{K}_{\mathrm{p}}\boldsymbol{e}(t)]+\boldsymbol{N}(\boldsymbol{q}(t),\dot{\boldsymbol{q}}(t)) \tag{5.2.5}$$

5.2.2　PI 控制器

图 5.2 是控制信号如式(5.2.6)所示的 PI 控制器框图。

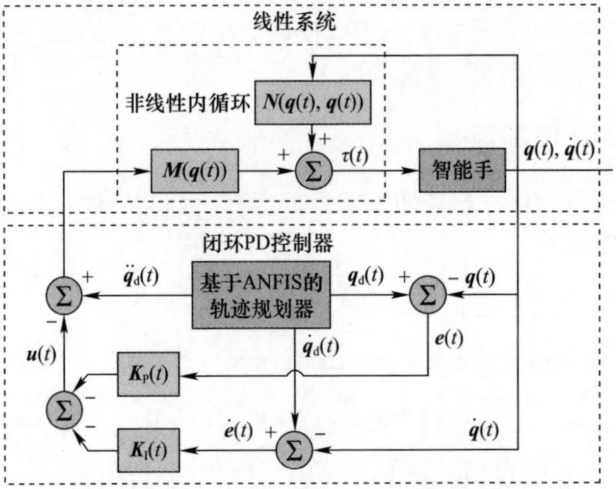

图 5.2　机器人手的融合 PI 控制器框图

$$u(t) = -K_p e(t) - K_I \int e(t)\mathrm{d}t \tag{5.2.6}$$

式中:K_I 是对角积分增益矩阵。定义如下:

$$\dot{\epsilon}(t) = e(t) \tag{5.2.7}$$

于是得到

$$\tau(t) = M(q(t))[\ddot{q}(t) + K_p e(t) + K_I \int e(t)\mathrm{d}t] + N(q(t),\dot{q}(t)) \tag{5.2.8}$$

5.2.3 PID 控制器

图 5.3 是 PID 控制器框图,其控制信号如下:

$$u(t) = -K_p e(t) - K_I \int e(t)\mathrm{d}t - K_D \dot{e}(t) \tag{5.2.9}$$

然后将(5.1.13) 改写为

$$\tau(t) = M(q(t))[\ddot{q}_d(t) + K_p e(t) + K_I \int e(t)\mathrm{d}t + K_D \dot{e}(t)] + N(q(t),\dot{q}(t)) \tag{5.2.10}$$

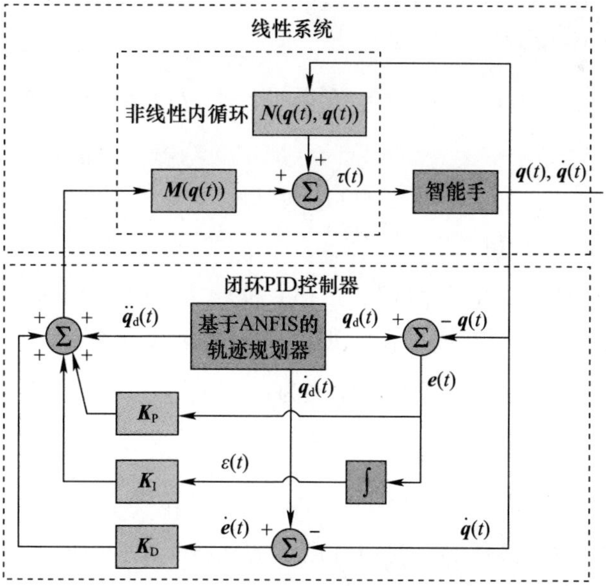

图 5.3　机械手的融合 PID 控制器框图

5.3 最优控制器

5.3.1 最优调节

最优是日常生活中非常理想的特征。最优控制的主要目标是确定控制信号,使其在运动过程中满足某些物理约束,同时对选定的性能标准(性能指标或成本函数)求取极值(极大值和极小值)[6]。

最优控制问题的制定需要如下条件:

(1) 被控过程的数学描述(或模型)(通常以状态变量形式表示);

(2) 规范的性能指标;

(3) 边界条件和物理约束状态或控制的一个说明。

5.3.2 带跟踪系统的线性二次型最优控制

从原始非线性系统(3.3.16)得到线性系统(5.1.14)时,没有状态空间的变换。进一步说,用于原始非线性系统(3.3.16)的控制器的困难设计已经转换为线性系统(5.1.15)控制器的简单设计。如果选择控制函数使线性系统(5.1.14)稳定并使跟踪误差为零,则由系统(5.1.13)给出的非线性转矩控制律将命令机械手系统(3.3.16)遵循所需的轨迹 $\boldsymbol{q}_{\mathrm{d}}(t)$。基于系统(5.1.13)给出的,初始机械手系统(3.3.16)变为

$$\boldsymbol{M}(\boldsymbol{q}(t))\ddot{\boldsymbol{q}}(t)+\boldsymbol{N}(\boldsymbol{q}(t),\dot{\boldsymbol{q}}(t))=\boldsymbol{M}(\boldsymbol{q}(t))[\ddot{\boldsymbol{q}}_{\mathrm{d}}(t)-\boldsymbol{u}(t)]+\boldsymbol{N}(\boldsymbol{q}(t),\dot{\boldsymbol{q}}(t))$$

即

$$\ddot{\boldsymbol{e}}(t)=\boldsymbol{u}(t) \tag{5.3.1}$$

这就是线性系统(5.1.14)。

我们的目标是控制线性系统(5.1.15),使状态变量 $\boldsymbol{x}(t)=[\boldsymbol{q}'(t)\ \dot{\boldsymbol{q}}'(t)]'$ 在最小控制能量的时间间隔 $[t_0,t_{\mathrm{f}}]$ 期间尽可能近地跟踪所需的输出 $\boldsymbol{z}(t)=[\boldsymbol{q}_{\mathrm{d}}'(t)\ \dot{\boldsymbol{q}}_{\mathrm{d}}'(t)]'$。为此,将误差矢量定义为

$$\boldsymbol{e}(t)=\boldsymbol{z}(t)-\boldsymbol{x}(t) \tag{5.3.2}$$

并选择性能指标 J[6]为

$$J=\frac{1}{2}\boldsymbol{e}'(t_{\mathrm{f}})\boldsymbol{F}(t_{\mathrm{f}})\boldsymbol{e}(t_{\mathrm{f}})+\frac{1}{2}\int_{t_0}^{t_{\mathrm{f}}}[\boldsymbol{e}'(t)\boldsymbol{Q}\boldsymbol{e}(t)+\boldsymbol{u}'(t)\boldsymbol{R}\boldsymbol{u}(t)]\mathrm{d}t \tag{5.3.3}$$

假设 $\boldsymbol{F}(t_{\mathrm{f}})$ 和 $\boldsymbol{Q}$ 是对称的半正定矩阵,$\boldsymbol{R}$ 是对称的正定矩阵。利用庞特里

亚金最小值原理[6]求解出矩阵微分黎卡提方程(DRE):

$$\dot{\boldsymbol{P}}(t) = -\boldsymbol{P}(t)\boldsymbol{A} - \boldsymbol{A}'\boldsymbol{P}(t) + \boldsymbol{P}(t)\boldsymbol{B}\boldsymbol{R}^{-1}\boldsymbol{B}'\boldsymbol{P}(t) - \boldsymbol{Q} \tag{5.3.4}$$

其中最终状态为 $\boldsymbol{P}(t_f) = \boldsymbol{F}(t_f)$。非齐次矢量微分方程为

$$\dot{\boldsymbol{g}}(t) = -[\boldsymbol{A} - \boldsymbol{B}\boldsymbol{R}^{-1}\boldsymbol{B}'\boldsymbol{P}(t)]'\boldsymbol{g}(t) - \boldsymbol{Q}\boldsymbol{z}(t) \tag{5.3.5}$$

其中最终状态为 $\boldsymbol{g}(t_f) = \boldsymbol{F}(t_f)\boldsymbol{z}(t_f)$。然后从中求解最优状态 $\dot{\boldsymbol{x}}^*(t)$,即

$$\dot{\boldsymbol{x}}^*(t) = [\boldsymbol{A} - \boldsymbol{B}\boldsymbol{R}^{-1}\boldsymbol{B}'\boldsymbol{P}(t)]\boldsymbol{x}^*(t) + \boldsymbol{B}\boldsymbol{R}^{-1}\boldsymbol{B}'\boldsymbol{g}(t) \tag{5.3.6}$$

其中初始状态为 $\boldsymbol{x}(t_0)$,最优控制 $\boldsymbol{u}^*(t)$ 由下式计算:

$$\boldsymbol{u}^*(t) = -\boldsymbol{R}^{-1}\boldsymbol{B}'\boldsymbol{P}(t)\boldsymbol{x}^*(t) + \boldsymbol{R}^{-1}\boldsymbol{B}'\boldsymbol{g}(t) \tag{5.3.7}$$

最终,最佳所需扭矩 $\boldsymbol{\tau}^*(t)$ 由下式获得,即

$$\boldsymbol{\tau}^*(t) = \boldsymbol{M}(\boldsymbol{q}(t))\boldsymbol{u}^*(t) + \boldsymbol{N}(\boldsymbol{q}(t), \dot{\boldsymbol{q}}(t)) \tag{5.3.8}$$

综上所述,图 5.4 显示了机械手在有限时间下线性二次型最优控制器跟踪系统的框图。使用反馈线性化技术将非线性动力学转换为线性。然后通过庞特里亚金最小值原理实现闭环有限时间线性二次型最优控制器,以实现采用三次多项式跟踪所需的轨迹规划。分别通过求解矩阵微分黎卡提方程和具有边界条件的非齐次矢量微分方程计算 $\boldsymbol{P}(t)$ 和 $\boldsymbol{g}(t)$。最后,获得最优状态 $\boldsymbol{x}^*(t)$ 和最优控制 $\boldsymbol{u}^*(t)$ 以便计算所需扭矩 $\boldsymbol{\tau}^*(t)$。

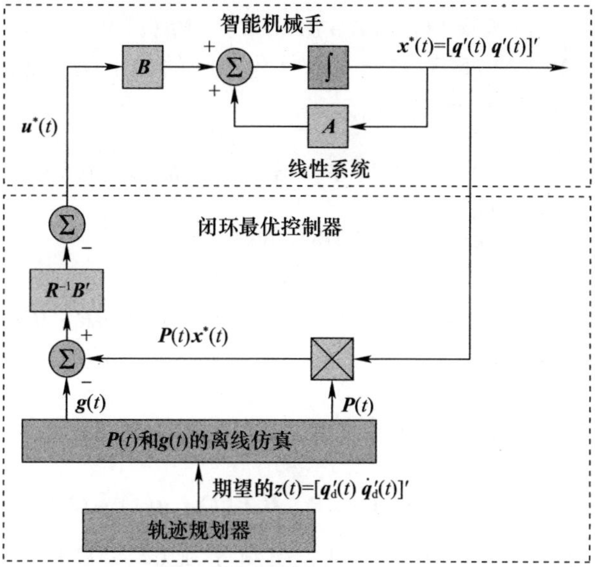

图 5.4　机械手的线性二次型最优控制器跟踪系统框图

5.3.3 带跟踪系统的改进最优控制

以前的工作[5,7]表明,初始的最优控制可以避免过冲和振荡问题,并获得比带有 GA 调谐功能的 PID 控制更好的结果[5,8],但这种最优控制方法在应用于机械手时需要一定的执行时间。为了提高初始最优控制器的性能,将性能指标 J[6]改为下式所示包含指数项的形式,即

$$\hat{J} = \frac{1}{2}\mathrm{e}^{2\alpha t_f}\boldsymbol{e}'(t_f)\boldsymbol{F}(t_f)\boldsymbol{e}(t_f) + \frac{1}{2}\int_{t_0}^{t_f}\mathrm{e}^{2\alpha t}[\boldsymbol{e}'(t)\boldsymbol{Q}\boldsymbol{e}(t) + \boldsymbol{u}'(t)\boldsymbol{R}\boldsymbol{u}(t)]\mathrm{d}t \tag{5.3.9}$$

式中:α 是一个正参数。我们需要找到在动态约束(5.1.15)下最小化新性能指标 $\hat{J}$(5.3.9)的最优控制。可以通过修改初始系统解决这个问题,因此可以推出以下转换形式:

$$\begin{cases}\hat{\boldsymbol{e}}(t) = \mathrm{e}^{\alpha t}\boldsymbol{e}(t), & \hat{\boldsymbol{z}}(t) = \mathrm{e}^{\alpha t}\boldsymbol{z}(t)\\ \hat{\boldsymbol{x}}(t) = \mathrm{e}^{\alpha t}\boldsymbol{x}(t), & \hat{\boldsymbol{u}}(t) = \mathrm{e}^{\alpha t}\boldsymbol{u}(t)\end{cases} \tag{5.3.10}$$

然后,使用转换方程(5.3.10),很容易获得新系统

$$\begin{aligned}\dot{\hat{\boldsymbol{x}}}(t) &= \frac{\mathrm{d}}{\mathrm{d}t}\{\mathrm{e}^{\alpha t}\boldsymbol{x}(t)\} = \alpha\mathrm{e}^{\alpha t}\boldsymbol{x}(t) + \mathrm{e}^{\alpha t}\dot{\boldsymbol{x}}(t)\\ &= \alpha\hat{\boldsymbol{x}}(t) + \mathrm{e}^{\alpha t}[\boldsymbol{A}\boldsymbol{x}(t) + \boldsymbol{B}\boldsymbol{u}(t)]\end{aligned}$$

$$\dot{\hat{\boldsymbol{x}}}(t) = (\boldsymbol{A} + \alpha\boldsymbol{I})\hat{\boldsymbol{x}}(t) + \boldsymbol{B}\hat{\boldsymbol{u}}(t) \tag{5.3.11}$$

考虑到式(5.3.11)和式(5.3.9)定义的优化系统的最小化实现,新的最优控制 $\hat{\boldsymbol{u}}^*(t)$与式(5.3.7)类似,如下式所示:

$$\hat{\boldsymbol{u}}^*(t) = -\boldsymbol{R}^{-1}\boldsymbol{B}'\hat{\boldsymbol{P}}(t)\hat{\boldsymbol{x}}^*(t) + \boldsymbol{R}^{-1}\boldsymbol{B}'\hat{\boldsymbol{g}}(t) \tag{5.3.12}$$

式中:矩阵 $\hat{\boldsymbol{P}}(t)$和矢量 $\hat{\boldsymbol{g}}(t)$分别是 DRE 的解,且

$$\dot{\hat{\boldsymbol{P}}}(t) = -\hat{\boldsymbol{P}}(t)(\boldsymbol{A} + \alpha\boldsymbol{I}) - (\boldsymbol{A}' + \alpha\boldsymbol{I})\hat{\boldsymbol{P}}(t) + \hat{\boldsymbol{P}}(t)\boldsymbol{B}\boldsymbol{R}^{-1}\boldsymbol{B}'\hat{\boldsymbol{P}}(t) - \boldsymbol{Q} \tag{5.3.13}$$

其中最终状态 $\hat{\boldsymbol{P}}(t_f) = \boldsymbol{F}(t_f)$,非齐次矢量微分方程为

$$\dot{\hat{\boldsymbol{g}}}(t) = -[\boldsymbol{A} + \alpha\boldsymbol{I} - \boldsymbol{B}\boldsymbol{R}^{-1}\boldsymbol{B}'\hat{\boldsymbol{P}}(t)]'\hat{\boldsymbol{g}}(t) - \boldsymbol{Q}\hat{z}(t) \tag{5.3.14}$$

其中最终状态 $\hat{\boldsymbol{g}}(t_f) = \boldsymbol{F}(t_f)\hat{z}(t_f)$。在新系统(5.3.11)中使用最优控制(式(5.3.12)),得到最优闭环系统

$$\dot{\hat{\boldsymbol{x}}}^*(t)=[\boldsymbol{A}+\alpha\boldsymbol{I}-\boldsymbol{B}\boldsymbol{R}^{-1}\boldsymbol{B}'\hat{\boldsymbol{P}}(t)]\hat{\boldsymbol{x}}^*(t)+\boldsymbol{B}\boldsymbol{R}^{-1}\boldsymbol{B}'\hat{\boldsymbol{g}}(t) \tag{5.3.15}$$

式中：初始状态为 $\hat{\boldsymbol{x}}(t_0)$。

因此，在新系统（5.3.15）中应用变换式（5.3.10），原始系统的最优控制（式(5.1.15)）和相关的性能指标（式(5.3.9)）由下式给出：

$$\begin{aligned}\boldsymbol{u}^*(t)&=\mathrm{e}^{-\alpha t}\hat{\boldsymbol{u}}^*(t)=-\mathrm{e}^{-\alpha t}\boldsymbol{R}^{-1}\boldsymbol{B}'[\hat{\boldsymbol{P}}(t)\hat{x}^*(t)-\hat{\boldsymbol{g}}(t)]\\&=-\boldsymbol{R}^{-1}\boldsymbol{B}'\hat{\boldsymbol{P}}(t)\boldsymbol{x}^*(t)+\mathrm{e}^{-\alpha t}\boldsymbol{R}^{-1}\boldsymbol{B}'\hat{\boldsymbol{g}}(t)\end{aligned} \tag{5.3.16}$$

有趣的是，这个期望的（原始的）最优控制具有相同的矩阵 DRE 解，即 $\hat{\boldsymbol{P}}(t)=\boldsymbol{P}(t)$，作为与式（5.3.16）和式（5.3.7）比较的新系统的最优控制，新系统的 $\hat{\boldsymbol{g}}(t)=\mathrm{e}^{\alpha t}\boldsymbol{g}(t)$。可以看到闭环最优控制系统（5.3.15）具有实部小于 $-\alpha$ 的特征值。换句话说，状态 $\boldsymbol{x}^*(t)$ 至少与 $\mathrm{e}^{-\alpha t}$[9] 接近零的速度一样快。

5.4 自适应控制器

自适应控制涉及修改控制器的控制律，以应对被控制系统的参数缓慢时变或不确定的情况。例如，当飞机飞行时，其质量将因燃料消耗而缓慢减少，此时就需要一种能够适应这种变化条件的控制律。自适应控制不同于鲁棒控制，因为它不需要关于这些不确定或时变参数界限的先验信息；鲁棒控制保证如果变化在给定的范围内，则控制律不需要改变，而自适应控制则与控制律的变化有关。

将自适应控制应用于机械手，即使控制器不知道机器人的质量，自适应控制器也能很好地工作。在初始误差之后，实际的关节角度与期望的关节角度紧密匹配。在自适应控制中，控制器动力学允许学习未知参数，从而性能随时间逐渐提高[2]。

跟踪误差 $\boldsymbol{e}(t)$ 和滤波跟踪误差 $\boldsymbol{r}(t)$ 定义为

$$\boldsymbol{e}(t)=\boldsymbol{q}_{\mathrm{d}}(t)-\boldsymbol{q}(t) \tag{5.4.1}$$

$$\boldsymbol{r}(t)=\dot{\boldsymbol{e}}(t)+\boldsymbol{\Lambda}\boldsymbol{e}(t) \tag{5.4.2}$$

式中：$\boldsymbol{q}_{\mathrm{d}}(t)$ 是关节的理想角度矢量；$\boldsymbol{q}(t)$ 是关节的实际角度矢量；$\boldsymbol{\Lambda}=\mathrm{diag}(\lambda_1,\lambda_2,\cdots,\lambda_n)$ 是正定对角增益矩阵。滤波误差（式(5.4.2)）确保整个系统的稳定性，因此跟踪误差（式(5.4.1)）是有界的。图 5.5 显示了自适应控制器的框图。跟踪误差 $\boldsymbol{e}(t)$ 由基于轨迹规划器得到的实际角度 $\boldsymbol{q}(t)$ 和期望角度 $\boldsymbol{q}_{\mathrm{d}}(t)$ 计算得到。然后，通过误差变化和参数 $\boldsymbol{\Lambda}$ 乘以误差来计算滤波的跟踪误差 $\boldsymbol{r}(t)$。机械手非线性系统所需扭矩 $\boldsymbol{\tau}(t)$ 由非线性项 $\boldsymbol{f}(t)$ 和增益 $\boldsymbol{K}_{\mathrm{D}}$ 乘以滤波后的跟踪误差计算得出。

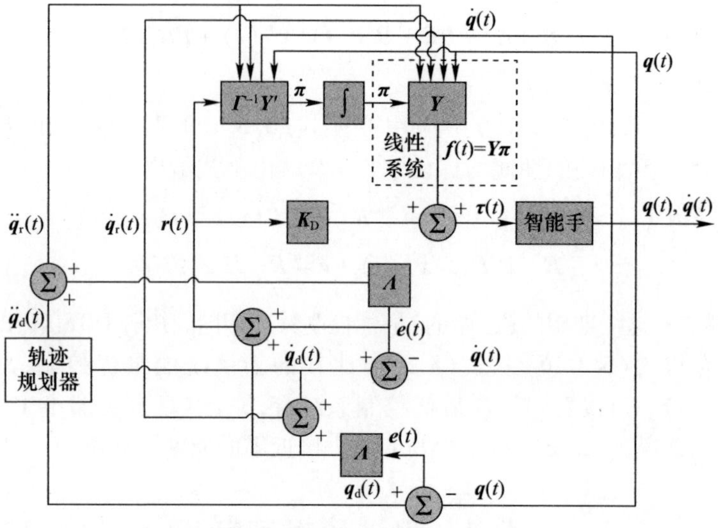

图 5.5　机械手自适应控制器的框图

将式(5.4.2)区分和替换成式(5.1.1)的过程给出了过滤误差 $\boldsymbol{r}(t)$ 的动态方程:

$$\boldsymbol{M}(\boldsymbol{q}(t))\dot{\boldsymbol{r}}(t) = -\boldsymbol{C}_{\mathrm{m}}(\boldsymbol{q}(t),\dot{\boldsymbol{q}}(t))\boldsymbol{r}(t) + \boldsymbol{f}(t) - \boldsymbol{\tau}(t) \tag{5.4.3}$$

式中:$\boldsymbol{C}(\boldsymbol{q}(t),\dot{\boldsymbol{q}}(t)) = \boldsymbol{C}_{\mathrm{m}}(\boldsymbol{q}(t),\dot{\boldsymbol{q}}(t))\dot{\boldsymbol{q}}(t)$;非线性项定义为

$$\begin{aligned}\boldsymbol{f}(t) &= \boldsymbol{M}(\boldsymbol{q}(t))(\ddot{\boldsymbol{q}}_{\mathrm{d}}(t) + \boldsymbol{\Lambda}\dot{\boldsymbol{e}}(t)) + \boldsymbol{G}(\boldsymbol{q}(t)) + \\ &\quad \boldsymbol{C}_{\mathrm{m}}(\boldsymbol{q}(t),\dot{\boldsymbol{q}}(t))(\dot{\boldsymbol{q}}_{\mathrm{d}}(t) + \boldsymbol{\Lambda}\boldsymbol{e}(t)) + \boldsymbol{\tau}_{\mathrm{dis}} \\ &= \boldsymbol{Y}\boldsymbol{\pi}\end{aligned} \tag{5.4.4}$$

式中:$\boldsymbol{\tau}_{\mathrm{dis}}$ 是未知的干扰;$\boldsymbol{Y}$ 是已知机器人函数的回归矩阵,是参数未知的矢量[10]。5. A[11] 附录给出了双关节拇指和三关节食指的回归矩阵 $\boldsymbol{Y}$ 和未知参数矢量。扭矩矢量 $\boldsymbol{\tau}(t)$ 由下式计算得到,即

$$\boldsymbol{\tau}(t) = \boldsymbol{f}(t) + \boldsymbol{K}_{\mathrm{D}}\boldsymbol{r}(t) \tag{5.4.5}$$

未知参数速率矢量 $\dot{\boldsymbol{\pi}}$ 更新为

$$\dot{\boldsymbol{\pi}} = \boldsymbol{\Gamma}^{-1}\boldsymbol{Y}'\boldsymbol{r}(t) \tag{5.4.6}$$

式中:$\boldsymbol{\Gamma}$ 是调整参数对角矩阵。

5.5　模拟结果与分析

本节介绍双关节拇指和三关节食指的智能机械手 PID 控制器、最优控制器和自适应控制器的模拟。然后介绍带有 14 个自由度的五指机械手的 PID 控制器、最优控制器和自适应控制器的模拟。

5.5.1 双关节拇指

表 5.1 中给出了与所需轨迹相关的各种参数[12]和为模拟选择的双关节拇指，目标方形物体的边长为 0.010m，如图 2.16 所示。PID 对角线系数 $K_P(t)$、$K_I(t)$ 和 $K_D(t)$ 是 100。对于最优控制系数，选择拇指的 $\boldsymbol{A}$、$\boldsymbol{B}$、$\boldsymbol{R}(t)$ 和 $\boldsymbol{Q}^t(t)$ 为

$$\boldsymbol{A} = \begin{bmatrix} \boldsymbol{0} & \boldsymbol{I} \\ \boldsymbol{0} & \boldsymbol{0} \end{bmatrix}, \quad \boldsymbol{B} = \begin{bmatrix} \boldsymbol{0} \\ \boldsymbol{I} \end{bmatrix}, \quad \boldsymbol{R}(t) = \frac{1}{30}\boldsymbol{I}, \quad \boldsymbol{Q}^t(t) = \begin{bmatrix} \boldsymbol{Q}_{11} & \boldsymbol{Q}_{12} \\ \boldsymbol{Q}_{12} & \boldsymbol{Q}_{22} \end{bmatrix}$$

$$\boldsymbol{Q}_{11} = \begin{bmatrix} 10 & 2 \\ 2 & 10 \end{bmatrix}, \quad \boldsymbol{Q}_{22} = \begin{bmatrix} 30 & 0 \\ 0 & 30 \end{bmatrix}, \quad \boldsymbol{Q}_{12} = \begin{bmatrix} -4 & 4 \\ 3 & -6 \end{bmatrix} \tag{5.5.1}$$

表 5.1 拇指的参数选择

参数	值	单位
时间(t_0,t_f)	0,20	s
期望的初始位置(X_0^t,Y_0^t)	0.035,0.060	m
期望的最终位置(X_f^t,Y_f^t)	0.0495,0.060	m
期望的初始速度($\dot{X}_0^t$,$\dot{Y}_0^t$)	0,0	m/s
期望的最终速度($\dot{X}_f^t$,$\dot{Y}_f^t$)	0,0	m/s
长度(L_1^t,L_2^t)	0.040,0.040	m
块(m_1^t,m_2^t)	0.043,0.031	kg
惯性(I_{zz1}^t,I_{zz2}^t)	$6.002^{\times 10^{-6}}$,$4.327^{\times 10^{-6}}$	$kg \cdot m^2$

图 5.6 显示了使用 PID 控制器的模拟结果，而图 5.7 显示了使用所提出的有限时间最优控制方法的模拟结果。可以清楚地看到，使用所提出的最优控制方法的仿真结果克服了过冲和振荡问题。下面提出一个应用于智能机械手的双关节拇指 PID 控制器和自适应控制器。图 5.8 显示了双关节拇指 PID 控制和自适应控制策略[13]。

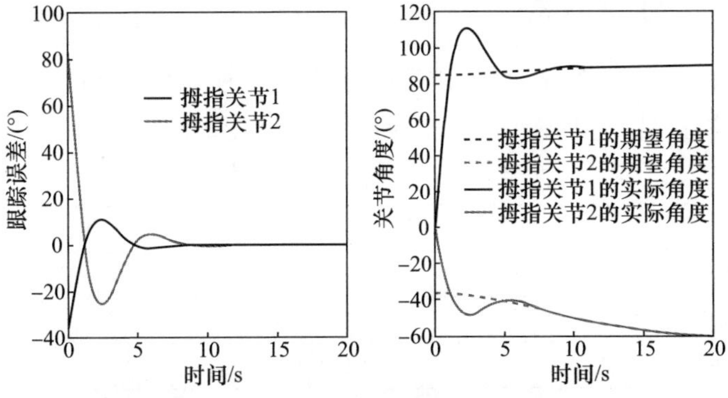

图 5.6 拇指 PID 控制器的跟踪误差和关节角度

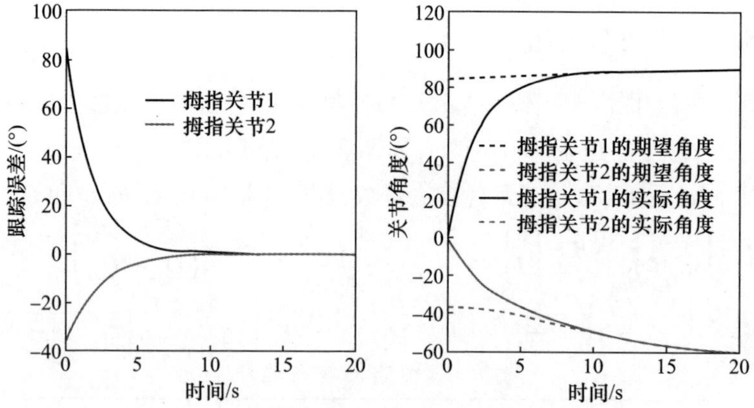

图 5.7 拇指最佳控制的跟踪误差和关节角度

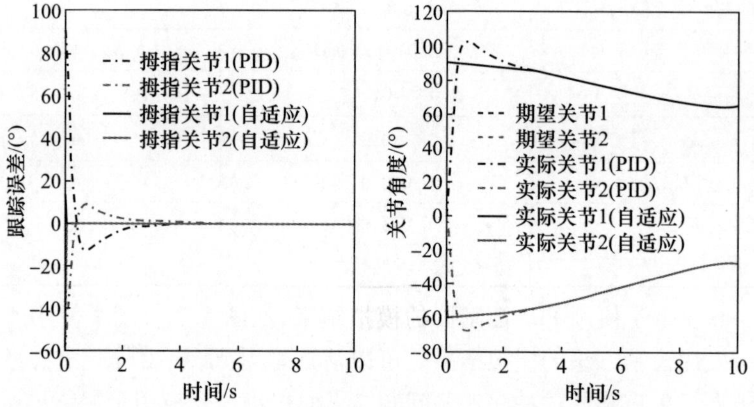

图 5.8 拇指 PID 和自适应控制器的跟踪误差和关节角度

5.5.2 三关节食指

类似地,表 5.2 中给出了与所需轨迹相关的各种参数[12]和为模拟选择的三关节食指,并且目标方形物体的边长为 0.010m,如图 2.16 所示。

表 5.2 食指的参数选择

参数	值	单位
时间(t_0,t_f)	0,20	s
期望的初始位置(X_0^t,Y_0^t)	0.065,0.080	m
期望的最终位置(X_f^t,Y_f^t)	0.010,0.060	m
期望的初始速度($\dot{X}_0^t$,$\dot{Y}_0^t$)	0,0	m/s

续表

参数	值	单位
期望的最终速度($\dot{X}_f^t$, $\dot{Y}_f^t$)	0,0	m/s
长度(L_1^i,L_2^i,L_3^i)	0.040,0.040,0.030	m
块(m_1^i,m_2^i,m_3^i)	0.045,0.025,0.017	kg
惯性($I_{zz1}^i,I_{zz2}^i,I_{zz3}^i$)	$9.375^{\times 10^{-6}}$, $3.333^{\times 10^{-6}}$, $1.125^{\times 10^{-6}}$	$kg \cdot m^2$
距离(d)	0.035	m

PID 对角线系数 $K_P(t)$、$K_I(t)$和 $K_D(t)$为 100,并且选择食指的最优控制系数 $\boldsymbol{A}$、$\boldsymbol{B}$、$R(t)$和 $Q^i(t)$为

$$\boldsymbol{A}=\begin{bmatrix}\boldsymbol{0} & \boldsymbol{I}\\ \boldsymbol{0} & \boldsymbol{0}\end{bmatrix},\quad \boldsymbol{B}=\begin{bmatrix}\boldsymbol{0}\\ \boldsymbol{I}\end{bmatrix},\quad \boldsymbol{R}(t)=\frac{1}{30}\boldsymbol{I},\quad \boldsymbol{Q}^i(t)=\begin{bmatrix}\boldsymbol{Q}_{11} & \boldsymbol{Q}_{12} & \boldsymbol{Q}_{13}\\ \boldsymbol{Q}_{12} & \boldsymbol{Q}_{22} & \boldsymbol{Q}_{23}\\ \boldsymbol{Q}_{13} & \boldsymbol{Q}_{23} & \boldsymbol{Q}_{33}\end{bmatrix}$$

$$\boldsymbol{Q}_{11}=\begin{bmatrix}10 & 2\\ 2 & 10\end{bmatrix},\quad \boldsymbol{Q}_{22}=\begin{bmatrix}30 & 0\\ 0 & 30\end{bmatrix},\quad \boldsymbol{Q}_{33}=\begin{bmatrix}20 & 1\\ 1 & 20\end{bmatrix}$$

$$\boldsymbol{Q}_{12}=\begin{bmatrix}-4 & 4\\ 3 & -6\end{bmatrix},\quad \boldsymbol{Q}_{13}=\begin{bmatrix}-4 & 4\\ 3 & -6\end{bmatrix},\quad \boldsymbol{Q}_{23}=\begin{bmatrix}-4 & 3\\ 4 & -6\end{bmatrix}$$

图 5.9 显示了使用 PID 控制器的模拟,而图 5.10 显示了使用所提出的有限时间最优控制方法的模拟。类似于拇指的结果,可以清楚地看到使用所提出的最优控制方法的结果可以克服过冲和振荡问题。

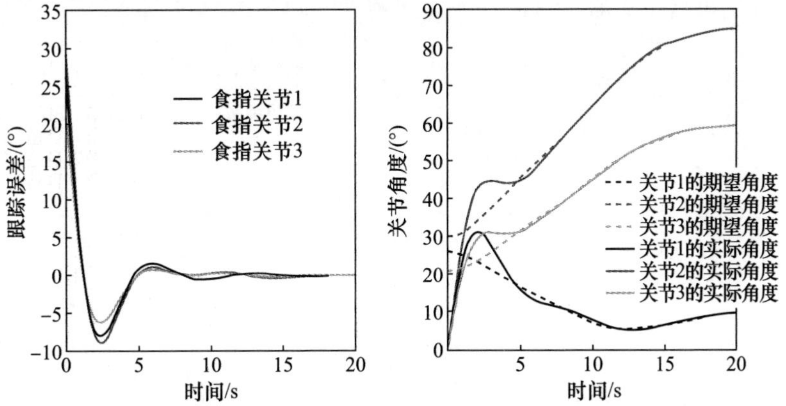

图 5.9　食指 PID 控制器的跟踪误差和关节角度

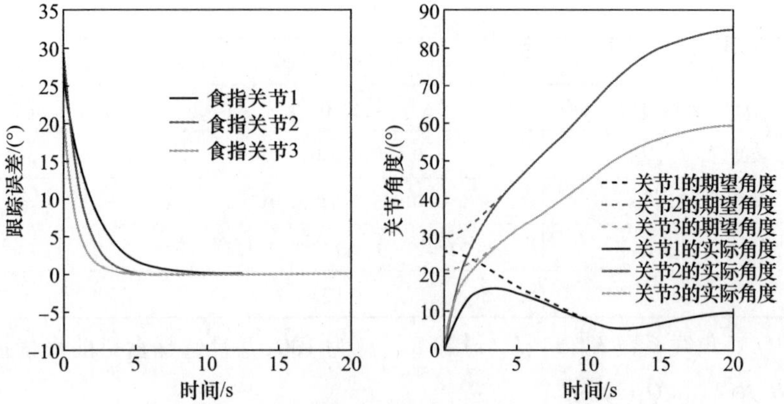

图 5.10　食指最佳控制的跟踪误差和关节角度

三关节食指的期望路径 2 如下：

$$\boldsymbol{q}_{\mathrm{d}}^{\mathrm{i}}=\left[A_1^{\mathrm{i}}\sin\left(\frac{2\pi}{T}\right)\quad A_2^{\mathrm{i}}\sin\left(\frac{2\pi}{T}\right)\quad 0.7A_2^{\mathrm{i}}\sin\left(\frac{2\pi}{T}\right)\right]' \tag{5.5.2}$$

与所需轨迹相关的各种参数[12]和为模拟选择的双关节拇指/三关节食指如下：

$A_1^{\mathrm{i}}=A_2^{\mathrm{i}}=0.1$；$T=2$（路径 2）；初始位置$(X^{\mathrm{t}},Y^{\mathrm{t}})=(0.035,0.060)$(m)，最终位置$(X^{\mathrm{t}},Y^{\mathrm{t}})=(0.0495,0.060)$(m)；初始和最终速度为零（路径 1）；关节 1、2 和 3 的长度分别为 0.040m、0.040m 和 0.030m。PID 对角线系数 $K_{\mathrm{P}}(t)$、$K_{\mathrm{I}}(t)$ 和 $K_{\mathrm{D}}(t)$ 为 100。对于自适应对角线系数 K_{D}、Λ 和 Γ 也选为 100。图 5.8 和图 5.11 显示了以前对[14-15]采取了 PID 控制器和自适应控制方法的双关节拇指和三关节食指所做的工作。可以清楚地看到，即使机械手受到干扰，使用所提出的融合自适应控制策略也克服了过冲和振荡问题[16]。

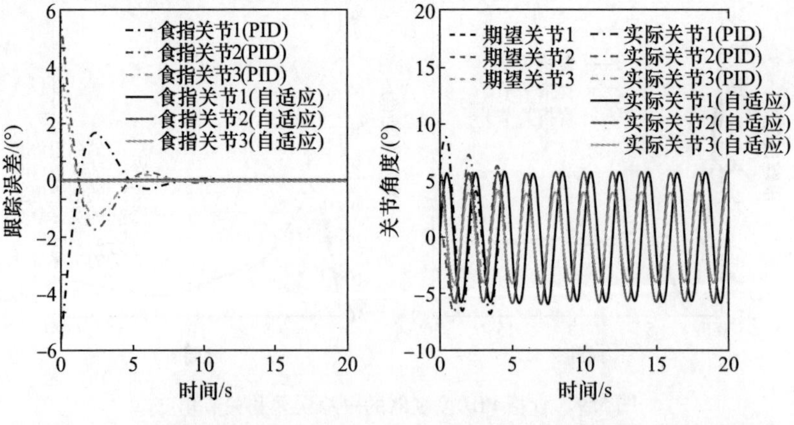

图 5.11　追踪手指路径 2 的 PID 和自适应控制器的误差和关节角度

5.5.3 三维五指机械手

下面对 14 自由度、五指智能机械手 PID 控制器和有限时间线性二次型最优控制器进行模拟。

5.5.3.1 PID 控制

双关节拇指/三关节食指[12]的参数与期望的轨迹有关。用于模拟的智能机械手的所有参数在表 5.3 中给出，目标矩形杆的宽度和长度分别为 0.010m 和 0.100m，如图 2.17 所示。表 5.4[11,17]中定义了全局坐标系和局部坐标系之间的相关参数。此外，假设所有关节都是半径(R)为 0.010m 的圆柱体，因此每个手指 $j(j=\mathrm{t},\mathrm{i},\mathrm{m},\tau,\mathrm{l})$ 上的每个关节 k 的惯性 I^j_{zzk} 计算式为

$$I^j_{zzk}=\frac{1}{4}m^j_kR^2+\frac{1}{3}m^j_kL^{j2}_k \tag{5.5.3}$$

所有初始实际角度均为零，所有 PID 对角线系数 $K_\mathrm{P}(t)$、$K_\mathrm{I}(t)$ 和 $K_\mathrm{D}(t)$ 均为 100。

表 5.3 智能机械手的参数选择

	参数	值	单位
拇指	时间(t_0,t_f)*	0,20	s
	期望的初始位置($X^\mathrm{t}_0,Y^\mathrm{t}_0$)**	0.035,0.060	m
	期望的最终位置($X^\mathrm{t}_\mathrm{f},Y^\mathrm{t}_\mathrm{f}$)**	0.0495,0.060	m
	期望的初始速度($\dot{X}^\mathrm{t}_0,\dot{Y}^\mathrm{t}_0$)*	0,0	m/s
	期望的最终速度($\dot{X}^\mathrm{t}_\mathrm{f},\dot{Y}^\mathrm{t}_\mathrm{f}$)*	0,0	m/s
	长度($L^\mathrm{t}_1,L^\mathrm{t}_2$)	0.040,0.040	m
	块($m^\mathrm{t}_1,m^\mathrm{t}_2$)	0.043,0.031	kg
食指	期望的初始位置($X^\mathrm{i}_0,Y^\mathrm{i}_0$)**	0.065,0.080	m
	期望的最终位置($X^\mathrm{i}_\mathrm{f},Y^\mathrm{i}_\mathrm{f}$)**	0.010,0.060	m
	长度($L^\mathrm{i}_1,L^\mathrm{i}_2,L^\mathrm{i}_3$)	0.040,0.040,0.030	m
	块($m^\mathrm{i}_1,m^\mathrm{i}_2,m^\mathrm{i}_3$)	0.045,0.025,0.017	kg
中指	期望的初始位置($X^\mathrm{m}_0,Y^\mathrm{m}_0$)**	0.065,0.080	m
	期望的最终位置($X^\mathrm{m}_\mathrm{f},Y^\mathrm{m}_\mathrm{f}$)**	0.005,0.060	m
	长度($L^\mathrm{m}_1,L^\mathrm{m}_2,L^\mathrm{m}_3$)	0.044,0.044,0.033	m
	块($m^\mathrm{m}_1,m^\mathrm{m}_2,m^\mathrm{m}_3$)	0.050,0.028,0.017	kg
无名指	期望的初始位置($X^\mathrm{r}_0,Y^\mathrm{r}_0$)**	0.065,0.080	m
	期望的最终位置($X^\mathrm{r}_\mathrm{f},Y^\mathrm{r}_\mathrm{f}$)**	0.010,0.060	m
	长度($L^\mathrm{r}_1,L^\mathrm{r}_2,L^\mathrm{r}_3$)	0.040,0.040,0.030	m
	块($m^\mathrm{r}_1,m^\mathrm{r}_2,m^\mathrm{r}_3$)	0.041,0.023,0.014	kg

续表

	参数	值	单位
小指	期望的初始位置(X_0^1, Y_0^1)**	0.055,0.080	m
	期望的最终位置(X_f^1, Y_f^1)**	0.020,0.060	m
	长度(L_1^1, L_2^1, L_3^1)	0.036,0.036,0.027	m
	块(m_1^1, m_2^1, m_3^1)	0.041,0.023,0.014	kg

注:*表示所有手指使用相同的参数;**表示所有参数都在本地坐标系中。

表 5.4 全局坐标系和本地坐标系之间的参数选择

参数	值	单位
α	90	(°)
β	45	(°)
d^{i}	(0.035,0.0)	m
d^{m}	(0.040,0,−0.020)	m
d^{r}	(0.035,0,−0.040)	m
d^{l}	(0.025,0,−0.060)	m

5.5.3.2 最优控制

所有手指的最优控制系数 $A,B,F(t_{\mathrm{f}})$ 和 $Q(t)$ 参数选择为

$$\boldsymbol{A}=\begin{bmatrix}\boldsymbol{0} & \boldsymbol{I}\\ \boldsymbol{0} & \boldsymbol{0}\end{bmatrix},\quad \boldsymbol{B}=\begin{bmatrix}\boldsymbol{0}\\ \boldsymbol{I}\end{bmatrix},\quad \boldsymbol{F}(t_{\mathrm{f}})=0,\quad \boldsymbol{R}(t)=\frac{1}{30}\boldsymbol{I}$$

$$\boldsymbol{Q}^{\mathrm{t}}(t)=\begin{bmatrix}\boldsymbol{Q}_{11} & \boldsymbol{Q}_{12}\\ \boldsymbol{Q}_{12} & \boldsymbol{Q}_{22}\end{bmatrix},\quad \boldsymbol{Q}^{j}(t)=\begin{bmatrix}\boldsymbol{Q}_{11} & \boldsymbol{Q}_{12} & \boldsymbol{Q}_{13}\\ \boldsymbol{Q}_{12} & \boldsymbol{Q}_{22} & \boldsymbol{Q}_{23}\\ \boldsymbol{Q}_{13} & \boldsymbol{Q}_{23} & \boldsymbol{Q}_{33}\end{bmatrix}$$

$$\boldsymbol{Q}_{11}=\begin{bmatrix}10 & 2\\ 2 & 10\end{bmatrix},\quad \boldsymbol{Q}_{22}=\begin{bmatrix}30 & 0\\ 0 & 30\end{bmatrix},\quad \boldsymbol{Q}_{33}=\begin{bmatrix}20 & 1\\ 1 & 20\end{bmatrix}$$

$$\boldsymbol{Q}_{12}=\begin{bmatrix}-4 & 4\\ 3 & -6\end{bmatrix},\quad \boldsymbol{Q}_{13}=\begin{bmatrix}-4 & 4\\ 3 & -6\end{bmatrix},\quad \boldsymbol{Q}_{23}=\begin{bmatrix}-4 & 3\\ 4 & -6\end{bmatrix}$$

式中:$j=\mathrm{i,m,r,l}$。通过将 $F(t_{\mathrm{f}})$ 作为零矩阵使用,可以忽略式(5.3.2)右侧的第一项。在这个例子中没有显著差异。

图 5.12 至图 5.16 显示了融合 PID 控制器和最优控制器的模拟结果。可以清楚地看出,融合最优控制器在抑制与融合 PID 控制器有关的过冲和瞬变方面具有优势。

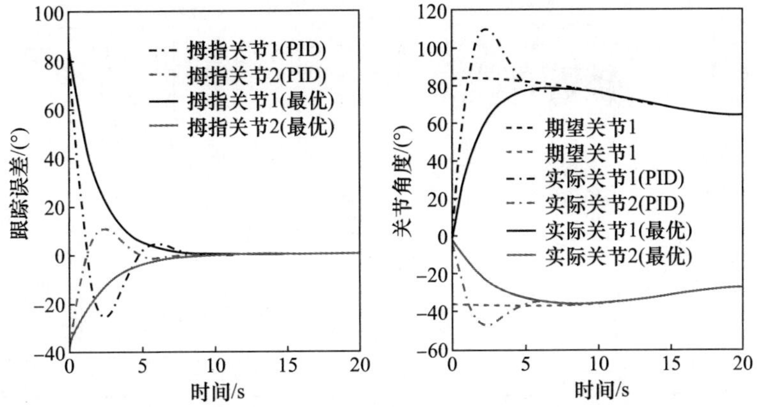

图 5.12　五指机械手拇指 PID 和最佳控制器的跟踪误差和关节角度

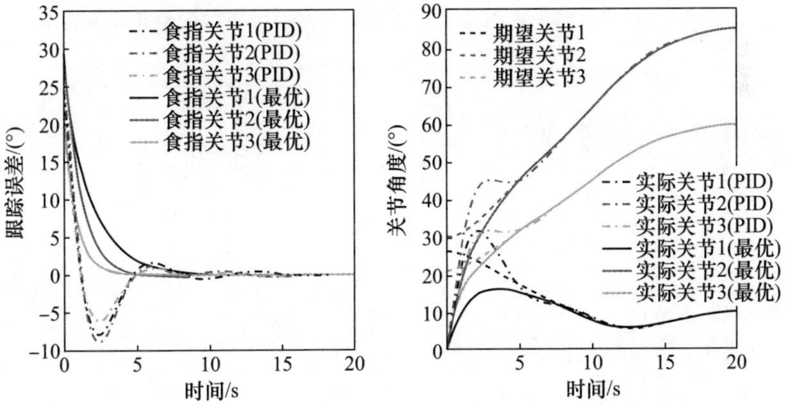

图 5.13　五指机械手食指 PID 和最佳控制器的跟踪误差和关节角度

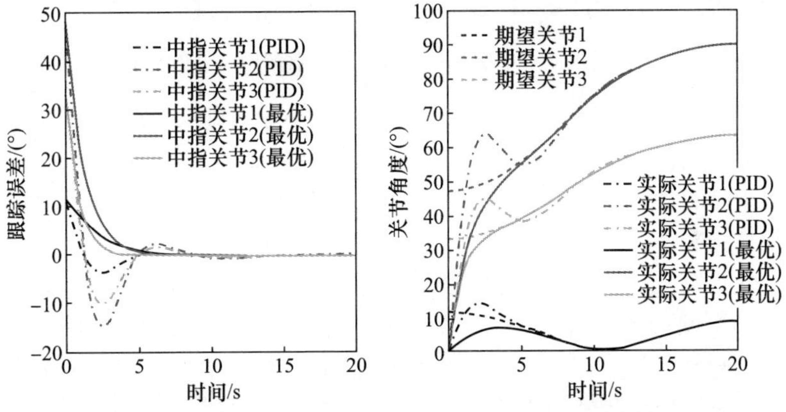

图 5.14　五指机械手中指 PID 和最佳控制器的跟踪误差和关节角度

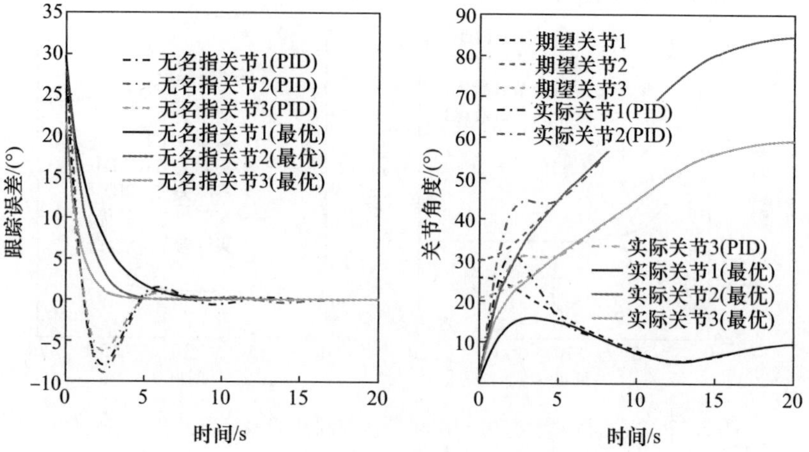

图 5.15　五指机械手无名指 PID 和最佳控制器的跟踪误差和关节角度

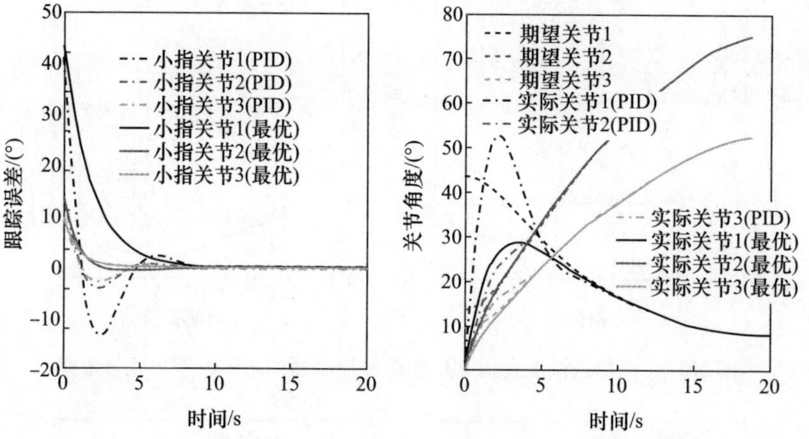

图 5.16　五指机械手小指 PID 和最佳控制器的跟踪误差和关节角度

图 5.17、图 5.19、图 5.21、图 5.23 和图 5.25 分别显示了所提出的五指智能机械手的拇指、食指、中指、无名指和小指的跟踪误差。图 5.18、图 5.20、图 5.22、图 5.24 和图 5.26 分别显示了所提出的五指智能机械手的拇指、食指、中指、无名指和小指的所需/实际角度。观察得到所有跟踪误差在 1s 内急剧下降并且在收敛后小于 1°,这提供了 14 自由度机械手的自适应控制器增强性能的证据。另一个观察结果表明,在收敛之后,所有三关节指状物表现出比双关节拇指更不稳定的误差,这表明自由度越大,在不知道所有指状物关节质量和惯性的情况下,自适应控制器的难度越大。

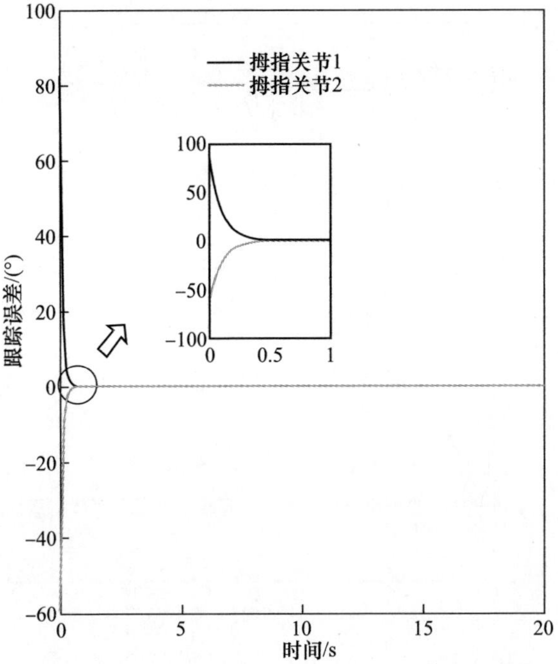

图 5. 17　双关节拇指自适应控制器的跟踪误差

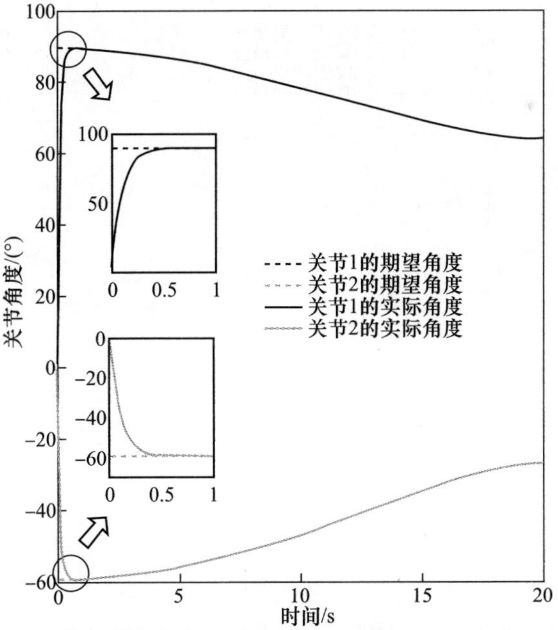

图 5. 18　双关节拇指自适应控制器的跟踪角度

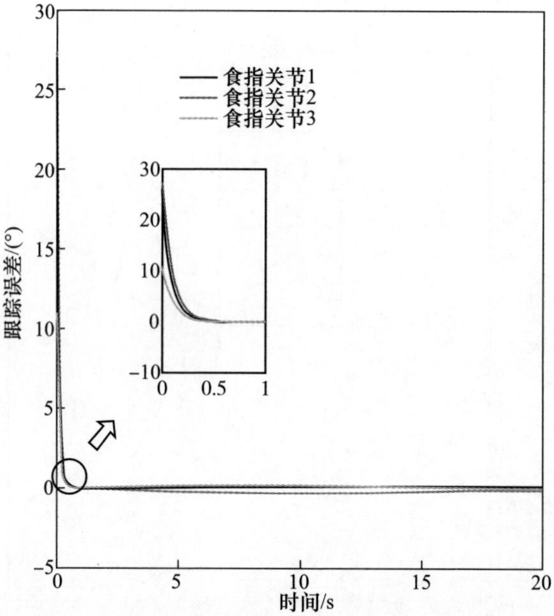

图 5.19　三关节食指自适应控制器的跟踪误差

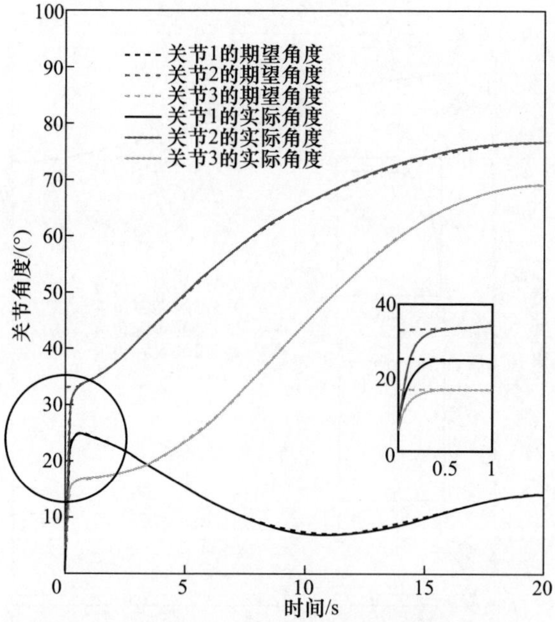

图 5.20　三关节食指自适应控制器的跟踪角度

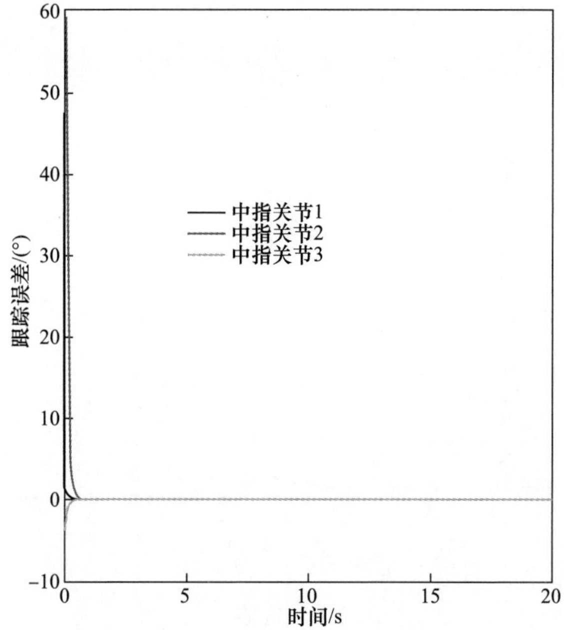

图 5.21　三关节中指自适应控制器的跟踪误差

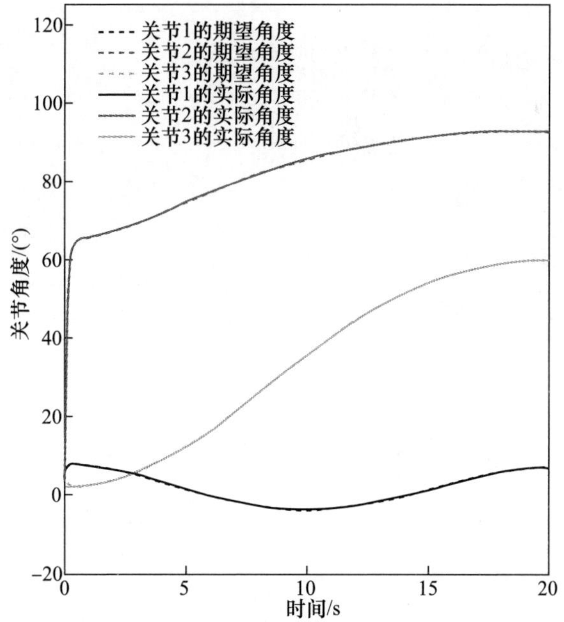

图 5.22　三关节中指自适应控制器的跟踪角度

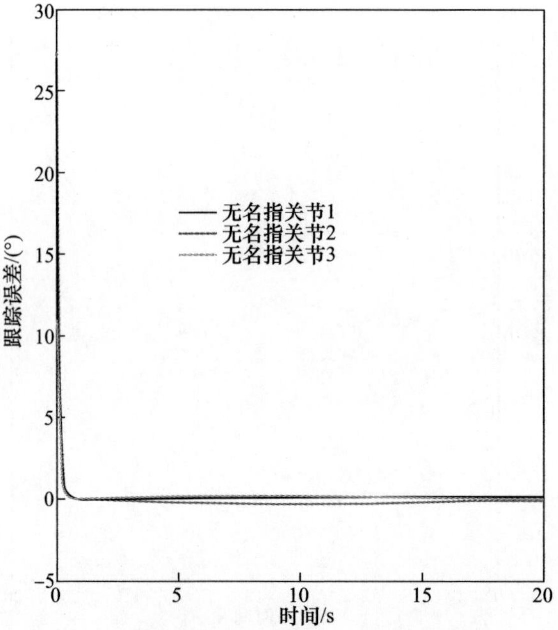

图 5. 23　三关节无名指自适应控制器的跟踪误差

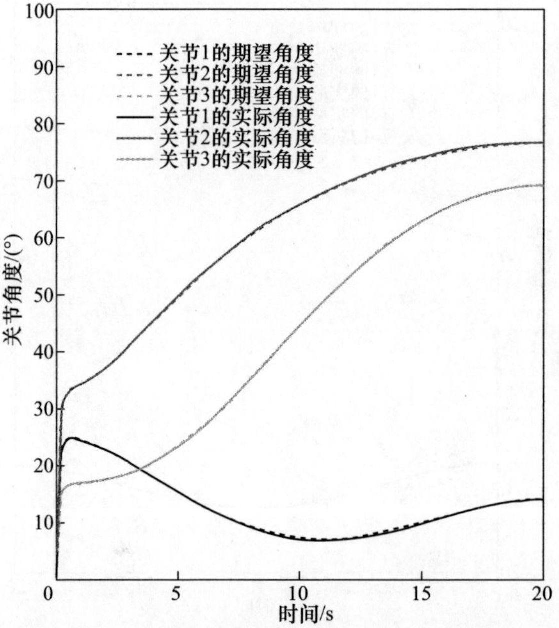

图 5. 24　三关节无名指自适应控制器的跟踪角度

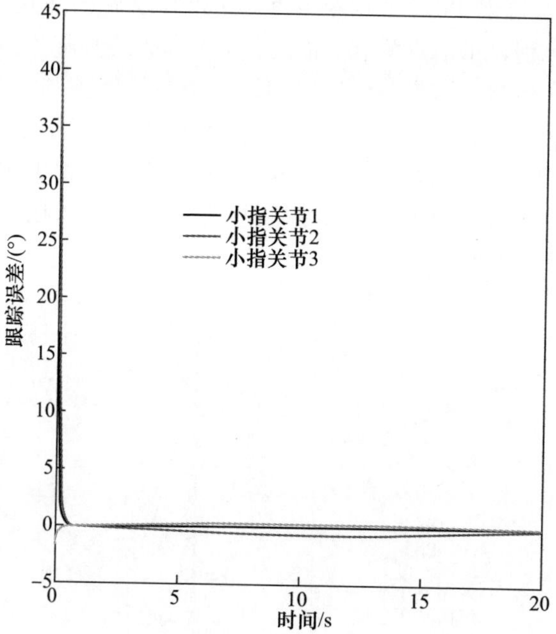

图 5.25　三关节小指自适应控制器的跟踪误差

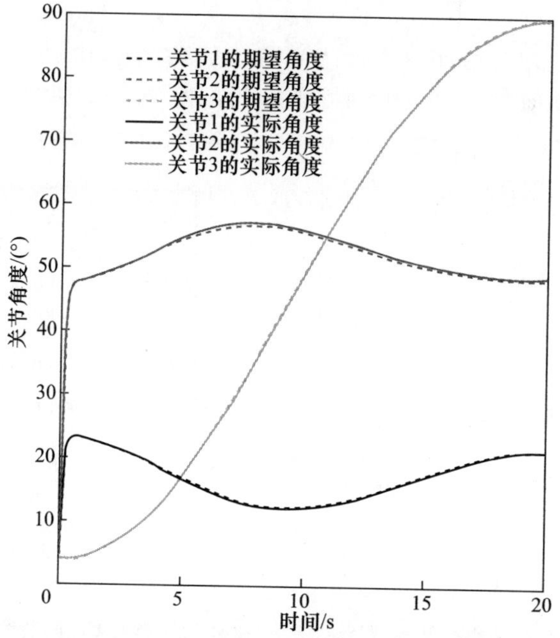

图 5.26　三关节小指自适应控制器的跟踪角度

为了比较 GA 调谐 PID（参见第 6.2 节）和修改后的最优控制器的性能，图 5.27和5.28 分别显示了双关节拇指的关节1 和2 的所需/实际角度和跟踪误差。GA 调谐 PID 控制显示了一个过冲问题。初始最优控制（$\alpha = 0°$）克服了这个问题，但是每个关节都至少需要 10s。随着参数从 1°增加到 10°，所提出的最优控制器的性能得到了提高。换句话说，当 $\alpha = 10°$ 时，收敛时间减少到大约 0.2s。对于三关节食指，GA 调谐 PID 控制不仅会引起过冲，还会引起振荡问题，如图 5.29（a）和 5.30 所示。

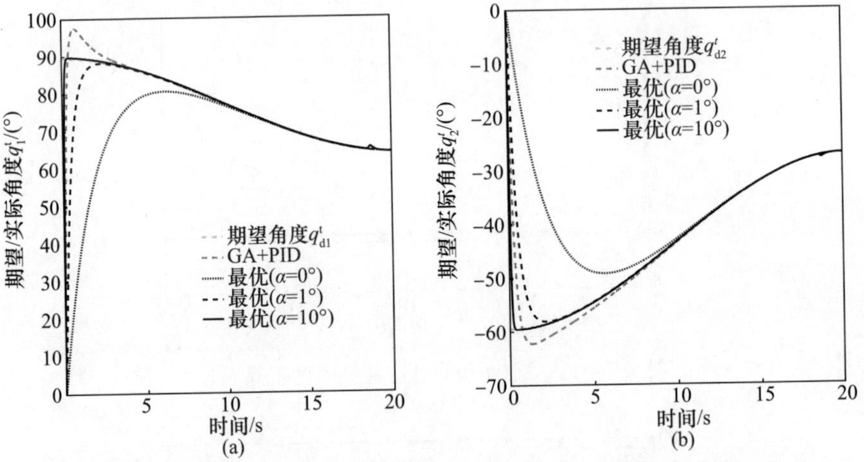

图 5.27　双关节拇指的期望/实际角位置（实际角度由 GA 调谐 PID 控制器和具有不同参数的修改后的最佳控制器调控，旨在跟踪所需的角度 q_{d1}^t 和 q_{d2}^t）

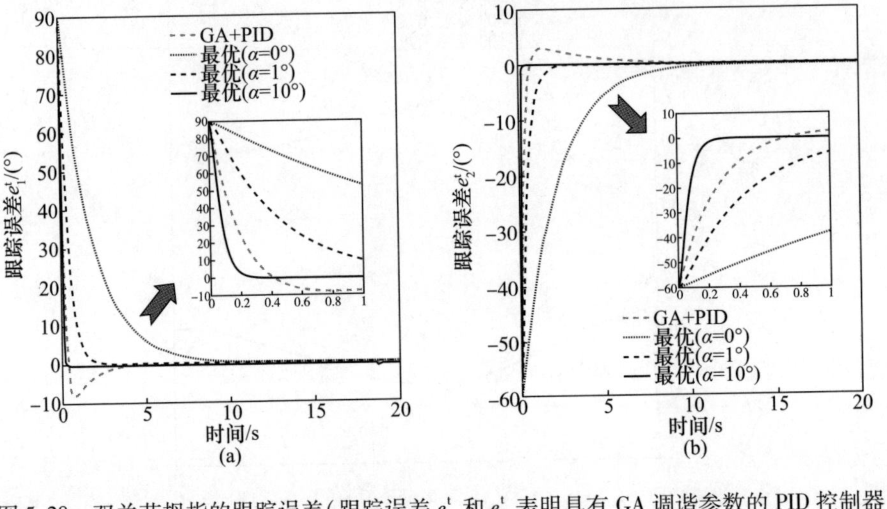

图 5.28　双关节拇指的跟踪误差（跟踪误差 e_1^t 和 e_2^t 表明具有 GA 调谐参数的 PID 控制器存在过冲问题，由提出的改进的最优控制器在 α 为 0°、1°和 10°时克服）

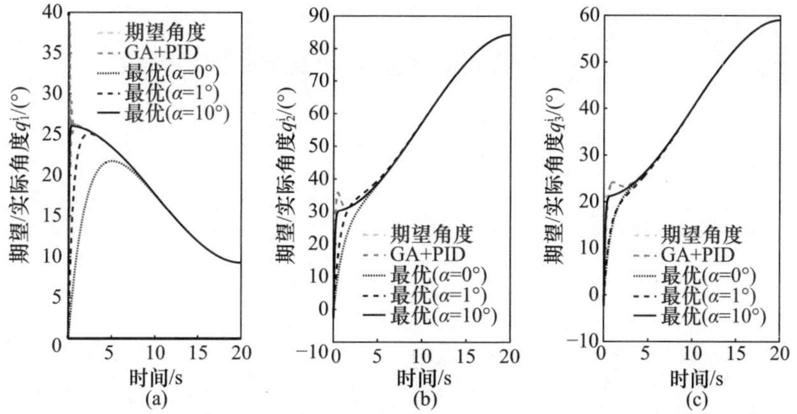

图 5.29　三关节食指的期望/实际角位置（实际角度 q_1^i、q_2^i 和 q_3^i，由 GA 调谐 PID 控制器和具有不同参数的改进的最优控制器调节，旨在分别跟踪所需的角度 q_{d1}^i、q_{d2}^i 和 q_{d3}^i）

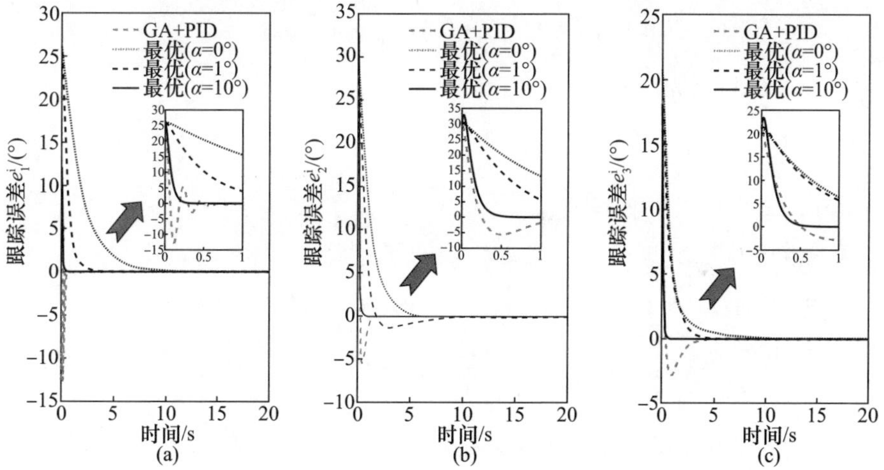

图 5.30　三关节食指的跟踪误差（跟踪误差 e_1^i、e_2^i 和 e_3^i 显示在参数 α 从 0°增加到 10°时改进的最优控制器比 GA 调谐 PID 控制器更快）

嵌入了指数项（α）的改进性能指标（$\hat{J}$）的最优控制也克服了过冲和振荡问题，并且随着 α 的增加获得了更快的收敛速度。我们也对其他三关节手指进行了类似的模拟实验。总之，这些数据表明，与 GA 调谐 PID 控制相比，改进的运动控制具有更高的精度和更快的收敛速度[9]。

5. A 附录：回归矩阵

在 5.4 节中，回归矩阵 $\boldsymbol{Y}^t$ 和双关节拇指的未知参数矢量 $\boldsymbol{\pi}^t$ 表示为

$$\boldsymbol{Y}^{\mathrm{t}}=\begin{bmatrix} Y_{11}^{\mathrm{t}} & Y_{12}^{\mathrm{t}} & Y_{13}^{\mathrm{t}} & Y_{14}^{\mathrm{t}} \\ Y_{21}^{\mathrm{t}} & Y_{22}^{\mathrm{t}} & Y_{23}^{\mathrm{t}} & Y_{24}^{\mathrm{t}} \end{bmatrix}$$

$$\boldsymbol{\pi}^{\mathrm{t}}=[m_1^{\mathrm{t}} \quad m_2^{\mathrm{t}} \quad m_{zz1}^{\mathrm{t}} \quad m_{zz2}^{\mathrm{t}}]$$

式中

$$Y_{11}^{\mathrm{t}}=l_1^{\mathrm{t}}l_1^{\mathrm{t}}(\dot{q}_{\mathrm{d}1}+\lambda_1\dot{e}_1)+gl_1^{\mathrm{t}}C_1$$

$$Y_{12}^{\mathrm{t}}=(2L_1^{\mathrm{t}}l_2^{\mathrm{t}}C_2+L_1^{\mathrm{t}}L_1^{\mathrm{t}}+l_2^{\mathrm{t}}l_2^{\mathrm{t}})(\ddot{q}_{\mathrm{d}1}+\lambda_1\dot{e}_1)+(L_1^{\mathrm{t}}l_2^{\mathrm{t}}C_2+l_2^{\mathrm{t}}l_2^{\mathrm{t}})(\ddot{q}_{\mathrm{d}2}+\lambda_2\dot{e}_2)-$$
$$L_1^{\mathrm{t}}l_2^{\mathrm{t}}S_2\dot{q}_2(\dot{q}_{\mathrm{d}1}+\lambda_1e_1)-L_1^{\mathrm{t}}l_2^{\mathrm{t}}S_2(\dot{q}_1+\dot{q}_2)(\dot{q}_{\mathrm{d}2}+\lambda_2e_2)+gl_2^{\mathrm{t}}C_{12}$$

$$Y_{13}^{\mathrm{t}}=\ddot{q}_{\mathrm{d}1}+\lambda_1\dot{e}_1$$

$$Y_{14}^{\mathrm{t}}=\ddot{q}_{\mathrm{d}1}+\lambda_1\dot{e}_1+\ddot{q}_{\mathrm{d}2}+\lambda_2\dot{e}_2$$

$$Y_{21}^{\mathrm{t}}=0$$

$$Y_{22}^{\mathrm{t}}=(L_1^{\mathrm{t}}l_2^{\mathrm{t}}C_2+l_2^{\mathrm{t}}l_2^{\mathrm{t}})(\ddot{q}_{\mathrm{d}1}+\lambda_1\dot{e}_1)+l_2^{\mathrm{t}}l_2^{\mathrm{t}}(\ddot{q}_{\mathrm{d}2}+\lambda_2\dot{e}_2)+$$
$$(L_1^{\mathrm{t}}l_2^{\mathrm{t}}S_2)\dot{q}_1(\dot{q}_{\mathrm{d}1}+\lambda_1e_1)-(L_1^{\mathrm{t}}l_2^{\mathrm{t}}S_2)\dot{q}_1(\dot{q}_{\mathrm{d}2}+\lambda_2e_2)+gl_2^{\mathrm{t}}C_{12}$$

$$Y_{23}^{\mathrm{t}}=0$$

$$Y_{24}^{\mathrm{t}}=\ddot{q}_{\mathrm{d}1}+\lambda_1\dot{e}_1+\ddot{q}_{\mathrm{d}2}+\lambda_2\dot{e}_2$$

$$C_1=\cos(q_1)$$

$$C_2=\cos(q_2)$$

$$S_2=\sin(q_2)$$

$$C_{12}=\cos(q_1+q_2)$$

类似地，三关节食指的回归矩阵和未知参数矢量可写为

$$\boldsymbol{Y}^{\mathrm{i}}=\begin{bmatrix} Y_{11}^{\mathrm{i}} & Y_{12}^{\mathrm{i}} & Y_{13}^{\mathrm{i}} & Y_{14}^{\mathrm{i}} & Y_{15}^{\mathrm{i}} & Y_{16}^{\mathrm{i}} \\ Y_{21}^{\mathrm{i}} & Y_{22}^{\mathrm{i}} & Y_{23}^{\mathrm{i}} & Y_{24}^{\mathrm{i}} & Y_{25}^{\mathrm{i}} & Y_{26}^{\mathrm{i}} \\ Y_{31}^{\mathrm{i}} & Y_{32}^{\mathrm{i}} & Y_{33}^{\mathrm{i}} & Y_{34}^{\mathrm{i}} & Y_{35}^{\mathrm{i}} & Y_{36}^{\mathrm{i}} \end{bmatrix}$$

$$\boldsymbol{\pi}^{\mathrm{i}}=[m_1^{\mathrm{i}} \quad m_2^{\mathrm{i}} \quad m_3^{\mathrm{i}} \quad I_{zz1}^{\mathrm{i}} \quad I_{zz2}^{\mathrm{i}} \quad I_{zz3}^{\mathrm{i}}]'$$

式中

$$Y_{11}^{\mathrm{i}}=l_1^{\mathrm{i}}l_1^{\mathrm{i}}(\ddot{q}_{\mathrm{d}1}+\lambda_1\dot{e}_1)+gl_1^{\mathrm{i}}C_1+gl_2^{\mathrm{i}}C_{12}$$

$$Y_{12}^{\mathrm{i}}=(2L_1^{\mathrm{i}}l_2^{\mathrm{i}}S_1S_{12}+2L_1^{\mathrm{i}}l_2^{\mathrm{i}}C_1C_{12}+L_1^{\mathrm{i}}L_1^{\mathrm{i}}+l_2^{\mathrm{i}}l_2^{\mathrm{i}})(\ddot{q}_{\mathrm{d}1}+\lambda_1\dot{e}_1)+$$

$$(L_1^i l_2^i S_1 S_{12} + L_1^i l_1^i C_1 C_{12} + l_2^i l_2^i)(\ddot{q}_{d2} + \lambda_2 \dot{e}_2) + gL_1 C_1 +$$
$$(L_1^i l_2^i S_1 C_{12} - L_1^i l_2^i C_1 S_{12})\dot{q}_1(\dot{q}_{d2} + \lambda_2 e_2) +$$
$$(L_1^i l_2^i S_1 C_{12} - L_1^i l_2^i C_1 S_{12})\dot{q}_2(\dot{q}_{d1} + \lambda_1 e_1) +$$
$$(L_1^i l_2^i S_1 C_{12} - L_1^i l_2^i C_1 S_{12})\dot{q}_2(\dot{q}_{d2} + \lambda_2 e_2)$$

$$Y_{13}^i = (2L_1^i L_2^i S_1 S_{12} + 2L_1^i L_2^i C_1 C_{12} + 2L_1^i l_3^i S_1 S_{123} + 2L_1^i l_3^i C_1 C_{123} +$$
$$2L_2^i l_3^i S_{12} S_{123} + 2L_2^i l_3^i C_{12} C_{123} + L_1^i L_1^i + L_2^i L_2^i + l_3^i l_3^i)(\ddot{q}_{d1} + \lambda_1 \dot{e}_1) +$$
$$(2L_2^i l_3^i S_{12} S_{123} + 2L_2^i l_3^i C_{12} C_{123} + L_1^i L_2^i S_1 S_{12} + L_1^i L_2^i C_1 C_{12} +$$
$$L_1^i l_3^i S_1 S_{123} + L_1^i l_3^i C_{12} C_{123} + L_2^i L_2^i + l_3^i l_3^i)(\ddot{q}_{d2} + \lambda_2 \dot{e}_2) +$$
$$(L_1^i l_3^i S_1 S_{123} + L_1^i l_3^i C_1 C_{123} + L_2^i l_3^i S_{12} S_{123} +$$
$$L_2^i l_3^i C_{12} C_{123} + l_3^i l_3^i)(\ddot{q}_{d3} + \lambda_3 \dot{e}_3) +$$
$$gL_1 C_1 + gL_2 C_{12} + gl_3 C_{123} +$$
$$(L_1^i L_2^i S_1 C_{12} - L_1^i L_2^i C_1 S_{12})\dot{q}_1(\dot{q}_{d2} + \lambda_2 e_2) +$$
$$(L_1^i L_2^i S_1 C_{12} - L_1^i L_2^i C_1 S_{12})\dot{q}_2(\dot{q}_{d1} + \lambda_1 e_1) +$$
$$(L_1^i l_3^i S_1 C_{123} - L_1^i l_3^i C_1 S_{123})\dot{q}_1(\dot{q}_{d2} + \lambda_2 e_2) +$$
$$(L_1^i l_3^i S_1 C_{123} - L_1^i l_3^i C_1 S_{123})\dot{q}_2(\dot{q}_{d1} + \lambda_1 e_1) +$$
$$(L_1^i l_3^i S_1 C_{123} - L_1^i l_3^i C_1 S_{123})\dot{q}_1(\dot{q}_{d3} + \lambda_3 e_3) +$$
$$(L_1^i l_3^i S_1 C_{123} - L_1^i l_3^i C_1 S_{123})\dot{q}_3(\dot{q}_{d1} + \lambda_1 e_1) +$$
$$(L_2^i l_3^i S_{12} C_{123} - L_2^i l_3^i C_{12} S_{123})\dot{q}_1(\dot{q}_{d3} + \lambda_3 e_3) +$$
$$(L_2^i l_3^i S_{12} C_{123} - L_2^i l_3^i C_{12} S_{123})\dot{q}_3(\dot{q}_{d1} + \lambda_1 e_1) +$$
$$(L_1^i l_3^i S_1 C_{123} - L_1^i l_3^i C_1 S_{123})\dot{q}_2(\dot{q}_{d3} + \lambda_3 e_3) +$$
$$(L_1^i l_3^i S_1 C_{123} - L_1^i l_3^i C_1 S_{123})\dot{q}_3(\dot{q}_{d2} + \lambda_2 e_2) +$$
$$(L_2^i l_3^i S_{12} C_{123} - L_2^i l_3^i C_{12} S_{123})\dot{q}_2(\dot{q}_{d3} + \lambda_3 e_3) +$$
$$(L_2^i l_3^i S_{12} C_{123} - L_2^i l_3^i C_{12} S_{123})\dot{q}_3(\dot{q}_{d2} + \lambda_2 e_2) +$$
$$(L_1^i L_2^i S_1 C_{12} - L_1^i L_2^i C_1 S_{12})\dot{q}_2(\dot{q}_{d2} + \lambda_2 e_2) +$$
$$(L_1^i l_3^i S_1 C_{123} - L_1^i l_3^i C_1 S_{123})\dot{q}_2(\dot{q}_{d2} + \lambda_2 e_2) +$$
$$(L_1^i l_3^i S_1 C_{123} - L_1^i l_3^i C_1 S_{123})\dot{q}_3(\dot{q}_{d3} + \lambda_3 e_3) +$$
$$(L_2^i l_3^i S_{12} C_{123} - L_2^i l_3^i C_{12} S_{123})\dot{q}_3(\dot{q}_{d3} + \lambda_3 e_3)$$

$$Y^{i}_{14}=\ddot{q}_{d1}+\lambda_1\dot{e}_1$$

$$Y^{i}_{15}=\ddot{q}_{d1}+\lambda_1\dot{e}_1+\ddot{q}_{d2}+\lambda_2\dot{e}_2$$

$$Y^{i}_{16}=\ddot{q}_{d1}+\lambda_1\dot{e}_1+\ddot{q}_{d2}+\lambda_2\dot{e}_2+\ddot{q}_{d3}+\lambda_3\dot{e}_3$$

$$Y^{i}_{21}=0$$

$$\begin{aligned}Y^{i}_{22}=&(L^{i}_1l^{i}_2S_1S_{12}+L^{i}_1l^{i}_2C_1C_{12}+l^{i}_2l^{i}_2)(\ddot{q}_{d1}+\lambda_1\dot{e}_1)+l^{i}_2l^{i}_2(\ddot{q}_{d2}+\lambda_2\dot{e}_2)+gl_2C_{12}+\\&(L^{i}_1l^{i}_2S_1C_{12}-L^{i}_1l^{i}_2C_1S_{12})\dot{q}_2(\dot{q}_{d1}+\lambda_1e_1)+\\&(L^{i}_1l^{i}_2C_1S_{12}-L^{i}_1l^{i}_2S_1C_{12})\dot{q}_1(\dot{q}_{d1}+\lambda_1e_1)\end{aligned}$$

$$\begin{aligned}Y^{i}_{23}=&(2L^{i}_2l^{i}_3S_{12}S_{123}+2L^{i}_2l^{i}_3C_{12}C_{123}+L^{i}_1L^{i}_2S_1S_{12}+L^{i}_1L^{i}_2C_1C_{12}+\\&L^{i}_1l^{i}_3S_1S_{123}+L^{i}_1l^{i}_3C_1C_{123}+L^{i}_2L^{i}_2+l^{i}_3l^{i}_3)(\ddot{q}_{d1}+\lambda_1\dot{e}_1)+\\&(2L^{i}_2l^{i}_3S_{12}S_{123}+2L^{i}_2l^{i}_3C_{12}C_{123}+L^{i}_2L^{i}_2+l^{i}_3l^{i}_3)(\ddot{q}_{d2}+\lambda_2\dot{e}_2)+\\&(L^{i}_2l^{i}_3S_{12}S_{123}+L^{i}_2l^{i}_3C_{12}C_{123}+l^{i}_3l^{i}_3)(\ddot{q}_{d3}+\lambda_3\dot{e}_3)+\\&gL_2C_{12}+gl_3C_{123}+(L^{i}_1L^{i}_2S_1C_{12}-L^{i}_1L^{i}_2C_1S_{12})\dot{q}_1(\dot{q}_{d2}+\lambda_2e_2)+\\&(L^{i}_1l^{i}_3S_1C_{123}-L^{i}_1l^{i}_3C_1S_{123})\dot{q}_1(\dot{q}_{d2}+\lambda_2e_2)+\\&(L^{i}_2l^{i}_3S_{12}C_{123}-L^{i}_2l^{i}_3C_{12}S_{123})\dot{q}_1(\dot{q}_{d3}+\lambda_3e_3)+\\&(L^{i}_2l^{i}_3S_{12}C_{123}-L^{i}_2l^{i}_3C_{12}S_{123})\dot{q}_3(\dot{q}_{d1}+\lambda_1e_1)+\\&(L^{i}_2l^{i}_3S_{12}C_{123}-L^{i}_2l^{i}_3C_{12}S_{123})\dot{q}_2(\dot{q}_{d3}+\lambda_3e_3)+\\&(L^{i}_2l^{i}_3S_{12}C_{123}-L^{i}_2l^{i}_3C_{12}S_{123})\dot{q}_3(\dot{q}_{d2}+\lambda_2e_2)+\\&(L^{i}_1L^{i}_2C_1S_{12}-L^{i}_1L^{i}_2S_1C_{12})\dot{q}_1(\dot{q}_{d1}+\lambda_1e_1)+\\&(L^{i}_1l^{i}_3C_1S_{123}-L^{i}_1l^{i}_3S_1C_{123})\dot{q}_1(\dot{q}_{d1}+\lambda_1e_1)+\\&(L^{i}_2l^{i}_3S_{12}C_{123}-L^{i}_2l^{i}_3C_{12}S_{123})\dot{q}_3(\dot{q}_{d3}+\lambda_3e_3)\end{aligned}$$

$$Y^{i}_{24}=0$$

$$Y^{i}_{25}=\ddot{q}_{d1}+\lambda_1\dot{e}_1+\ddot{q}_{d2}+\lambda_2\dot{e}_2$$

$$Y^{i}_{26}=\ddot{q}_{d1}+\lambda_1\dot{e}_1+\ddot{q}_{d2}+\lambda_2\dot{e}_2+\ddot{q}_{d3}+\lambda_3\dot{e}_3$$

$$Y^{i}_{31}=0$$

$$Y^{i}_{32}=0$$

$$\begin{aligned}Y^{i}_{33}=&(L^{i}_1l^{i}_3S_1S_{123}+L^{i}_1l^{i}_3C_1C_{123}+L^{i}_2l^{i}_3S_{12}S_{123}+\\&L^{i}_2l^{i}_3C_{12}C_{123}+l^{i}_3l^{i}_3)(\ddot{q}_{d1}+\lambda_1\dot{e}_1)+\end{aligned}$$

$$(L_2^{\mathrm{i}} l_3^{\mathrm{i}} S_{12} S_{123} + L_2^{\mathrm{i}} l_3^{\mathrm{i}} C_{12} C_{123} + l_3^{\mathrm{i}} l_3^{\mathrm{i}})(\ddot{q}_{\mathrm{d}2} + \lambda_2 \dot{e}_2) +$$
$$l_3^{\mathrm{i}} l_3^{\mathrm{i}} (\ddot{q}_{\mathrm{d}3} + \lambda_3 \dot{e}_3) + g l_3 C_{123} +$$
$$(L_2^{\mathrm{i}} l_3^{\mathrm{i}} C_{12} S_{123} - L_2^{\mathrm{i}} l_3^{\mathrm{i}} S_{12} C_{123}) \dot{q}_1 (\dot{q}_{\mathrm{d}2} + \lambda_2 e_2) +$$
$$(L_2^{\mathrm{i}} l_3^{\mathrm{i}} C_{12} S_{123} - L_2^{\mathrm{i}} l_3^{\mathrm{i}} S_{12} C_{123}) \dot{q}_2 (\dot{q}_{\mathrm{d}1} + \lambda_1 e_1) +$$
$$(L_1^{\mathrm{i}} l_3^{\mathrm{i}} S_1 C_{123} - L_1^{\mathrm{i}} l_3^{\mathrm{i}} C_1 S_{123}) \dot{q}_1 (\dot{q}_{\mathrm{d}3} + \lambda_3 e_3) +$$
$$(L_2^{\mathrm{i}} l_3^{\mathrm{i}} S_{12} C_{123} - L_2^{\mathrm{i}} l_3^{\mathrm{i}} C_{12} S_{123}) \dot{q}_1 (\dot{q}_{\mathrm{d}3} + \lambda_3 e_3) +$$
$$(L_2^{\mathrm{i}} l_3^{\mathrm{i}} S_{12} C_{123} - L_2^{\mathrm{i}} l_3^{\mathrm{i}} C_{12} S_{123}) \dot{q}_2 (\dot{q}_{\mathrm{d}3} + \lambda_3 e_3) +$$
$$(L_1^{\mathrm{i}} l_3^{\mathrm{i}} C_1 S_{123} - L_1^{\mathrm{i}} l_3^{\mathrm{i}} S_1 C_{123}) \dot{q}_1 (\dot{q}_{\mathrm{d}1} + \lambda_1 e_1) +$$
$$(L_2^{\mathrm{i}} l_3^{\mathrm{i}} C_{12} S_{123} - L_2^{\mathrm{i}} l_3^{\mathrm{i}} S_{12} C_{123}) \dot{q}_1 (\dot{q}_{\mathrm{d}1} + \lambda_1 e_1) +$$
$$(L_2^{\mathrm{i}} l_3^{\mathrm{i}} C_{12} S_{123} - L_2^{\mathrm{i}} l_3^{\mathrm{i}} S_{12} C_{123}) \dot{q}_2 (\dot{q}_{\mathrm{d}2} + \lambda_2 e_2)$$

$$Y_{34}^{\mathrm{i}} = 0$$
$$Y_{35}^{\mathrm{i}} = 0$$
$$Y_{36}^{\mathrm{i}} = \ddot{q}_{\mathrm{d}1} + \lambda_1 \dot{e}_1 + \ddot{q}_{\mathrm{d}2} + \lambda_2 \dot{e}_2 + \ddot{q}_{\mathrm{d}3} + \lambda_3 \dot{e}_3$$
$$C_1 = \cos(q_1)$$
$$C_{12} = \cos(q_1 + q_2)$$
$$C_{123} = \cos(q_1 + q_2 + q_3)$$
$$S_1 = \sin(q_1)$$
$$S_{12} = \sin(q_1 + q_2)$$
$$S_{123} = \sin(q_1 + q_2 + q_3)$$

参考文献

[1] M. J. Marquez. *Nonlinear Control Systems: Analysis and Design.* Wiley – Interscience, Hoboken, New Jersey, 2003.

[2] F. L. Lewis, D. M. Dawson, and C. T. Abdallah. *Robot Manipulators Control: Second Edition, Revised and Expanded.* Marcel Dekker, Inc., New York, USA, 2004.

[3] F. L. Lewis, S. Jagannathan, and A. Yesildirak. *Neural Network Control of Robot Manipulators and Non – Linear Systems.* CRC, 1998.

[4] R. Kelly, V. Santibanez, and A. Loria. *Control of Robot Manipulators in Joint Space.* Springer, New York, USA, 2005.

[5] C. – H. Chen, D. S. Naidu, A. Perez – Gracia, and M. P. Schoen. "A hybrid control strategy for five – fingered smart prosthetic hand," in *Joint 48th IEEE Conference on Decision and Control (CDC) and 28th Chinese Control Conference (CCC)*, pp. 5102 – 5107, Shanghai, P. R. China, December 16 – 18, 2009.

[6] D. S. Naidu. *Optimal Control Systems.* CRC Press, a Division of Taylor & Francis, Boca Raton, FL and London, UK, 2003 (A vastly expanded and updated version of this book, is under preparation for publication in 2017).

[7] C. - H. Chen, D. S. Naidu, A. Perez - Gracia, and M. P. Schoen. "A hybrid optimal control strategy for a smart prosthetic hand," in *Proceedings of the ASME 2009 Dynamic Systems and Control Conference (DSCC)*, Hollywood, California, USA, October 12 - 14, 2009 (No. DSCC2009 - 2507).

[8] C. - H. Chen and D. S. Naidu. "Hybrid genetic algorithm PID control for a five - fingered smart prosthetic hand," in *Proceedings of the Sixth International Conference on Circuits, Systems and Signals (CSS' 11)*, pp. 57 - 63, Vouliagmeni Beach, Athens, Greece, March 7 - 9, 2012.

[9] C. - H. Chen and D. S. Naidu. A modified optimal control strategy for a five - finger robotic hand. *International Journal of Robotics and Automation Technology*, 1(1): 3 - 10, November 2014.

[10] F. L. Lewis, S. Jagannathan, and A. Yesildirek. *Neural Network Control of Robotic Manipulators and Nonlinear Systems.* Taylor & Francis, London, UK, 1999.

[11] C. - H. Chen, D. S. Naidu, and M. P. Schoen. Adaptive control for a five - fingered prosthetic hand with unknown mass and inertia. *World Scientific and Engineering Academy and Society (WSEAS) Journal on Systems*, 10(5): 148 - 161, May 2011.

[12] S. Arimoto. Control Theory of Multi - fingered Hands: *A Modeling and Analytical - Mechanics Approach for Dexterity and Intelligence.* Springer - Verlag, London, UK, 2008.

[13] C. - H. Chen, D. S. Naidu, A. Perez - Gracia, and M. P. Schoen. "A hybrid adaptive control strategy for a smart prosthetic hand," in *The 31st Annual International Conference of the IEEE Engineering Medicine and Biology Society (EMBS)*, pp. 5056 - 5059, Minneapolis, Minnesota, USA, September 2 - 6, 2009.

[14] C. - H. Chen, K. W. Bosworth, M. P. Schoen, S. E. Bearden, D. S. Naidu, and A. Perez - Gracia. "A study of particle swarm optimization on leukocyte adhesion molecules and control strategies for smart prosthetic hand," in *2008 IEEE Swarm Intelligence Symposium (IEEE SIS08)*, St. Louis, Missouri, USA, September 21 - 23, 2008.

[15] C. - H. Chen, D. S. Naidu, A. Perez - Gracia, and M. P. Schoen. "Fusion of hard and soft control techniques for prosthetic hand," in *Proceedings of the International Association of Science and Technology* for *Development (IASTED) International Conference on Intelligent Systems and Control (ISC 2008)*, pp. 120 - 125, Orlando, Florida, USA, November 16 - 18, 2008.

[16] C. - H. Chen, D. S. Naidu, and M. P. Schoen. "An adaptive control strategy for a five - fingered prosthetic hand" in *The 14th world Scientific and Engineering Academy and Society (WSEAS) International Conference on Systems, Latest Trends on Systems (Volume II)*, pp. 405 - 410, Corfu Island, Greece, July 22 - 24, 2010.

[17] C. - H. Chen and D. S. Naidu. Hybrid control strategies for a five - finger robotic hand. *Biomedical Signal Processing and Control*, 8(4): 382 - 390, July 2013.

第 6 章　硬控制策略和软控制策略的融合 Ⅱ

本章介绍硬控制(HC)和软控制(SC)策略的融合以提高控制机械手/假肢的性能,这是单独使用硬控制或软控制策略所无法实现的。6.1 节介绍基于模糊逻辑的比例 - 微分(PD)融合控制策略,6.2 节介绍基于遗传算法的比例 - 积分 - 微分(PID)融合控制策略。

6.1　基于模糊逻辑的 PD 融合控制策略

图 6.1 是基于融合模糊逻辑的 PD 控制器框图,用来控制所提出的五指机器人,其控制输入信号为

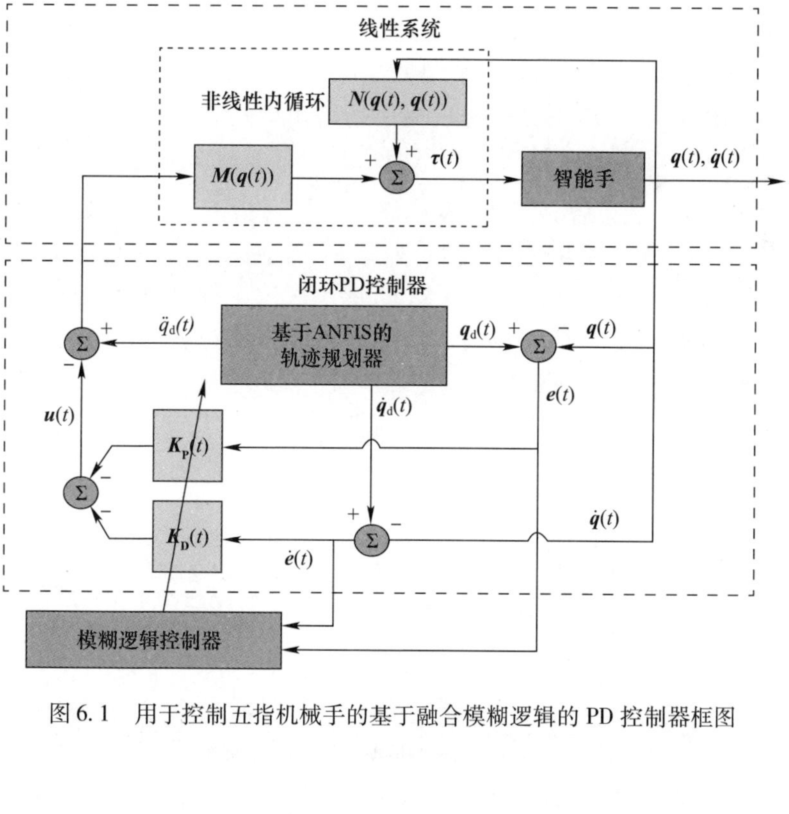

图 6.1　用于控制五指机械手的基于融合模糊逻辑的 PD 控制器框图

$$\boldsymbol{u}(t) = -\boldsymbol{K}_{\mathrm{P}}(t)\boldsymbol{e}(t) - \boldsymbol{K}_{\mathrm{D}}(t)\dot{\boldsymbol{e}}(t) \tag{6.1.1}$$

并且比例项 $\boldsymbol{K}_{\mathrm{P}}(t)$ 和微分项 $\boldsymbol{K}_{\mathrm{D}}(t)$ 的对角增益矩阵随时间 t 变化,将式(5.1.17)重新写成

$$\boldsymbol{\tau}(t) = \boldsymbol{M}(\boldsymbol{q}(t))[\ddot{\boldsymbol{q}}_{\mathrm{d}}(t) + \boldsymbol{K}_{\mathrm{P}}(t)\boldsymbol{e}(t) + \boldsymbol{K}_{\mathrm{D}}(t)\dot{\boldsymbol{e}}(t) + \boldsymbol{N}(\boldsymbol{q}(t), \dot{\boldsymbol{q}}(t))] \tag{6.1.2}$$

然后由于闭环 PD 控制器具有简单的 min - max 结构[1],我们使用 Mamdani 模糊推理系统来调整其时变参数 $\boldsymbol{K}_{\mathrm{P}}(t)$ 和 $\boldsymbol{K}_{\mathrm{D}}(t)$。值得注意的是,S. J. Ovaska 将 12 种智能融合系统与硬控制和软控制在结构融合类别中进行了比较,并给出了使用的融合等级的简要定义,这是描述硬控制和软控制结构之间特定连接强度的定性测量:低、中、高、很高[2]。在他的结构融合类别中,当前提出的基于融合模糊逻辑的 PD 控制器的融合等级被分类为非常高。

Mamdani 模型是由 Mamdani 和 Assilian 在 1975 年提出的用来控制一台蒸汽机和锅炉组合的机器[1]。该模型使用的语言控制规则是从经验丰富的人类操作员处得到的。从那时起,Mamdani 模糊系统就因其简单的 min - max 结构而成为常用的模糊推理方法。

图 6.2(a)是 Mamdani 模糊逻辑控制器框图。来自机器手的精确输入(误差 $\boldsymbol{e}(t)$ 和误差变化率 $\dot{\boldsymbol{e}}(t)$)由图 6.2(b)所示的 7 个三角隶属函数对其进行模糊化。隶属函数的数量通过不断尝试和失败实验得到。如果这个数量低于 7,那么模型的输出就不能够完美遵循机械手的输出。然后通过人类知识推理和表 6.1所列的 7×7 逻辑条件规则在模糊系统中对模糊输入进行并行处理。例如,"IF"(误差 $\boldsymbol{e}(t)$ 是负小,NS)和(误差变化率 $\dot{\boldsymbol{e}}(t)$ 为正中,PM)"THEN"($\boldsymbol{K}_{\mathrm{P}}(t)$ 为中等大,ML)。模糊输出之后被图 6.2(b)所示的另外 7 个三角隶属函数去模糊化以产生精确输出 $\boldsymbol{K}_{\mathrm{P}}(t)$。图 6.2(c)展示了这个模糊推理系统的输出面 $\boldsymbol{K}_{\mathrm{P}}(t)$。相同的,另一个精确输出 $\boldsymbol{K}_{\mathrm{D}}(t)$ 通过同样的过程计算得到。自适应 $\boldsymbol{K}_{\mathrm{P}}(t)$ 和 $\boldsymbol{K}_{\mathrm{D}}(t)$ 参数在闭环比例 - 微分控制器中得到使用。

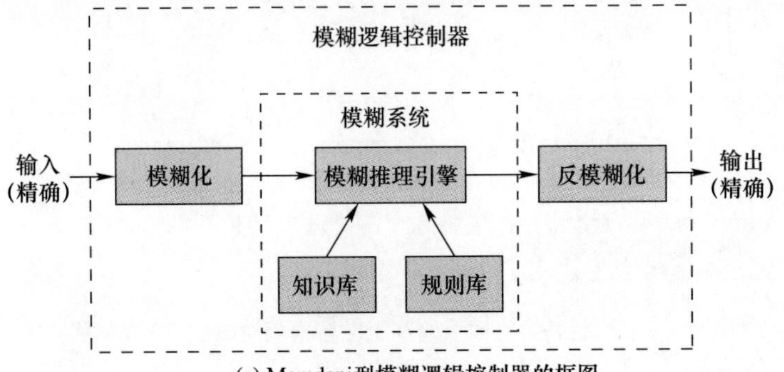

(a) Mamdani型模糊逻辑控制器的框图

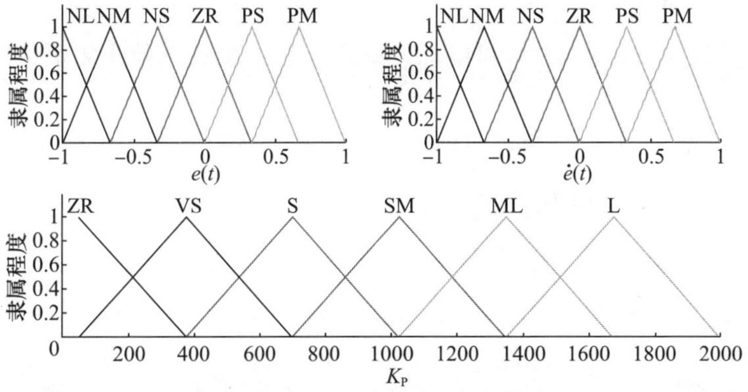

(b) 使用7个三角形隶属函数的两个输入（上）和一个输出（下）的所有隶属函数

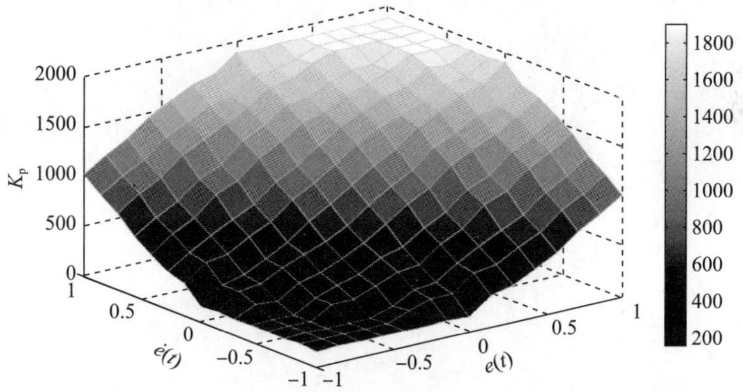

(c) 基于两个清晰输入和49个模糊逻辑知识规则的模糊推理系统的输出面

图 6.2　模糊逻辑控制器的结构

表 6.1　一套 7 ×7 模糊逻辑“IF – THEN”规则库

$\dot{e}\backslash e$	NL	NM	NS	ZR	PS	PM	PL
NL	ZR	ZR	ZR	ZR	VS	S	SM
NM	ZR	ZR	ZR	VS	S	SM	ML
NS	ZR	ZR	VS	S	SM	ML	L
ZR	ZR	VS	S	SM	ML	L	VL
PS	VS	S	SM	ML	L	VL	VL
PM	S	SM	ML	L	VL	VL	VL
PL	SM	ML	L	VL	VL	VL	VL

注:N,负数;P,正数;ZR,零;L,大;M,中等;S,小;V,非常。

总的来说,通过自适应神经模糊推理系统(ANFIS)轨迹规划器计算出的实际角度 $\boldsymbol{q}(t)$ 和期望角度 $\boldsymbol{q}_{\mathrm{d}}(t)$ 可以计算得到误差 $\boldsymbol{e}(t)$ 和误差变化率 $\dot{\boldsymbol{e}}(t)$。然后,模糊逻辑控制器调整闭环 PD 控制器的所有时变参数 $\boldsymbol{K}_{\mathrm{P}}(t)$ 和 $\boldsymbol{K}_{\mathrm{D}}(t)$,以便通过控制输入信号 $\boldsymbol{u}(t)$ 计算出机械手非线性系统所需要的转矩 $\boldsymbol{\tau}(t)$。

6.1.1 模拟结果和讨论

为了比较硬控制、软控制以及所提出的融合控制策略的精度和能耗,提出使用 PD 和 PID 控制器以及模糊推理系统的模拟来调整 PD 控制器,用于图 2.11 所示的具有 14 自由度的五指机械手以抓住矩形杆。双关节拇指/三关节手指的参数与期望的轨迹有关[3]。所有用于机械手模拟的参数均在表 5.3 中给出,并且目标矩形杆的边长和长度分别为 0.010m 和 0.100m。全局坐标和局部坐标之间相关参数的定义在图 2.12 中,它们的值在表 5.4 中给出。当拇指和四指进行伸展/屈曲运动时,指尖的工作范围受最大关节角度所限制。参考反向运动学,拇指的第一和第二关节角度被限制在[0°,90°]和[-80°,0°]范围内。其他四指的第一、第二和第三关节角度分别被约束在[0°,90°],[0°,110°]和[0°,80°]范围内[4]。

此外,假设所有手指的每个关节都是半径为 0.01m($R=0.010$m)的圆柱体,则所有手指 $j(j=\mathrm{t,i,m,r,l})$ 的每个关节 $k(k=1\sim3)$ 的惯性 I_{zzk}^{j} 的计算公式为

$$I_{zzk}^{j}=\frac{1}{4}m_k^jR^2+\frac{1}{3}m_k^jL_k^{j2} \tag{6.1.3}$$

实际角度的所有初始值都是零,并且单独的 PD 和 PID 控制器对角线系数 $\boldsymbol{K}_{\mathrm{P}}(t)$、$\boldsymbol{K}_{\mathrm{I}}(t)$ 和 $\boldsymbol{K}_{\mathrm{D}}(t)$ 被任意选择为 100。在导出的动态和控制模型中,选择参数($\boldsymbol{K}_{\mathrm{P}}(t)$ 和 $\boldsymbol{K}_{\mathrm{D}}(t)$)之后计算控制信号 $\boldsymbol{u}(t)$ 和转矩 $\boldsymbol{\tau}(t)$。图 6.3 是使用 PD(虚线)、PID(点线)、模糊逻辑(短虚线)以及模糊逻辑和 PD(实线)控制器的融合双关节拇指的关节 1(顶部)和 2(底部)的跟踪误差 $e_1^{\mathrm{t}}(t)$ 和 $e_2^{\mathrm{t}}(t)$ 以及期望/实际角度 $q_1^{\mathrm{t}}(t)$ 和 $q_2^{\mathrm{t}}(t)$。PD 和模糊逻辑控制器的跟踪误差在 5s 内收敛到零而没有超调,但是 PID 控制器存在超调和振荡需要更长的时间收敛到零(大约 10s)。所提出的使用参数 $\boldsymbol{K}_{\mathrm{P}}(t)\in[50,1000]$ 和 $\boldsymbol{K}_{\mathrm{D}}(t)\in[50,500]$ 的模糊逻辑和 PD 的融合控制的收敛速度比单独使用 PD、PID 和模糊逻辑控制器快 5~10 倍。模糊逻辑控制器包括两个输入(误差和误差变化率)和一个输出(控制输入信号)。为了进一步研究参数范围是否影响跟踪误差,我们发现在 $\boldsymbol{K}_{\mathrm{P}}(t)\in[50,1000]$ 且没有额外的计算时间[5]时,参数范围越大,收敛速度越快。这些结果清楚地证明硬控制和软控制策略的融合优于单独的硬控制或软控制方法。

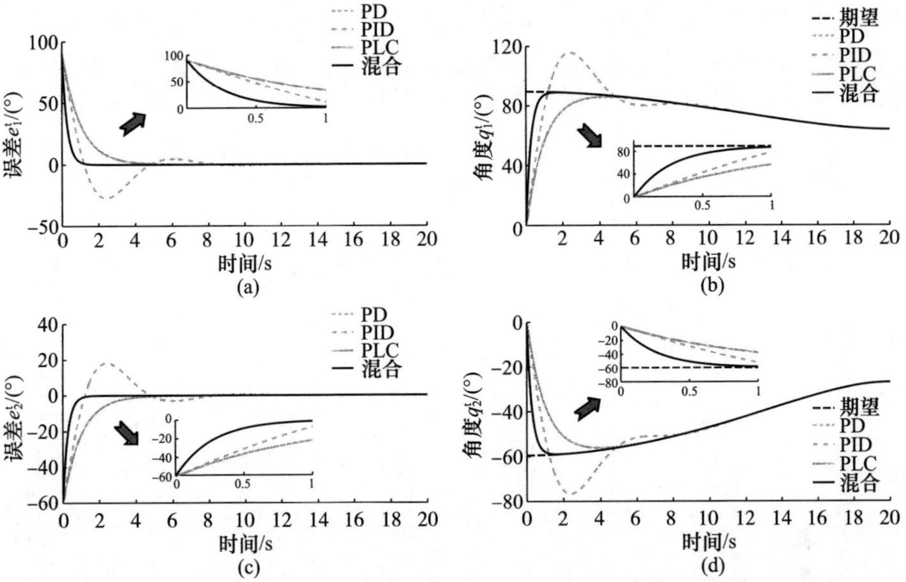

图 6.3　使用 PD(虚线)、PID(点线)、模糊逻辑(短虚线)及融合模糊逻辑和 PD(实线)控制器的双关节拇指的关节 1((a)、(b))和 2((c)、(d))的误差和期望/实际角度

双关节拇指的时变计算控制信号($u_1^t(t)$ 和 $u_2^t(t)$)和转矩($\tau_1^t(t)$ 和 $\tau_2^t(t)$)如图 6.4 所示,表明所提出的融合模糊逻辑的 PD 控制器比 PD、PID 和模糊逻辑控制器需要更多的转矩(能耗),以便获得收敛更快的跟踪误差。

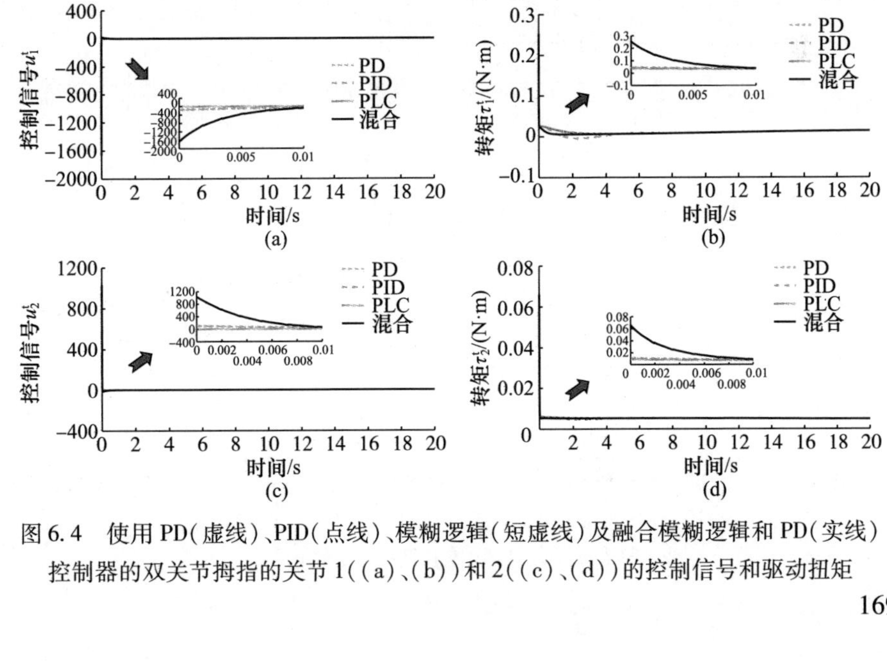

图 6.4　使用 PD(虚线)、PID(点线)、模糊逻辑(短虚线)及融合模糊逻辑和 PD(实线)控制器的双关节拇指的关节 1((a)、(b))和 2((c)、(d))的控制信号和驱动扭矩

对于三关节食指,图 6.5 是跟踪误差和所需/实际角度,图 6.6 分别是控制信号和扭矩。剩余的三关节手指(中指、无名指和小拇指)同样获得了相似的跟踪误差结果(数据未显示)。随着自由度的增加,PID 控制仍然显示出过冲和振荡的不良特征;PD 和模糊逻辑控制器都会降低精度和收敛速度。对于基于融合模糊逻辑的 PD 控制,它增加了所需的功率,但当自由度增加时,该融合控制器仍保持快速收敛和高精度。这些研究结果表明了所提出的提高精度的融合控制策略可能应用于危险环境的工业机器人、手术等。但是由于当前的电池容量受限,这种融合方法在机器人设备的临床应用仍存在限制。此限制可以通过实施硬控制和/或软控制方法来优化成本函数中嵌入的控制输入(性能指标)来解决[6]。

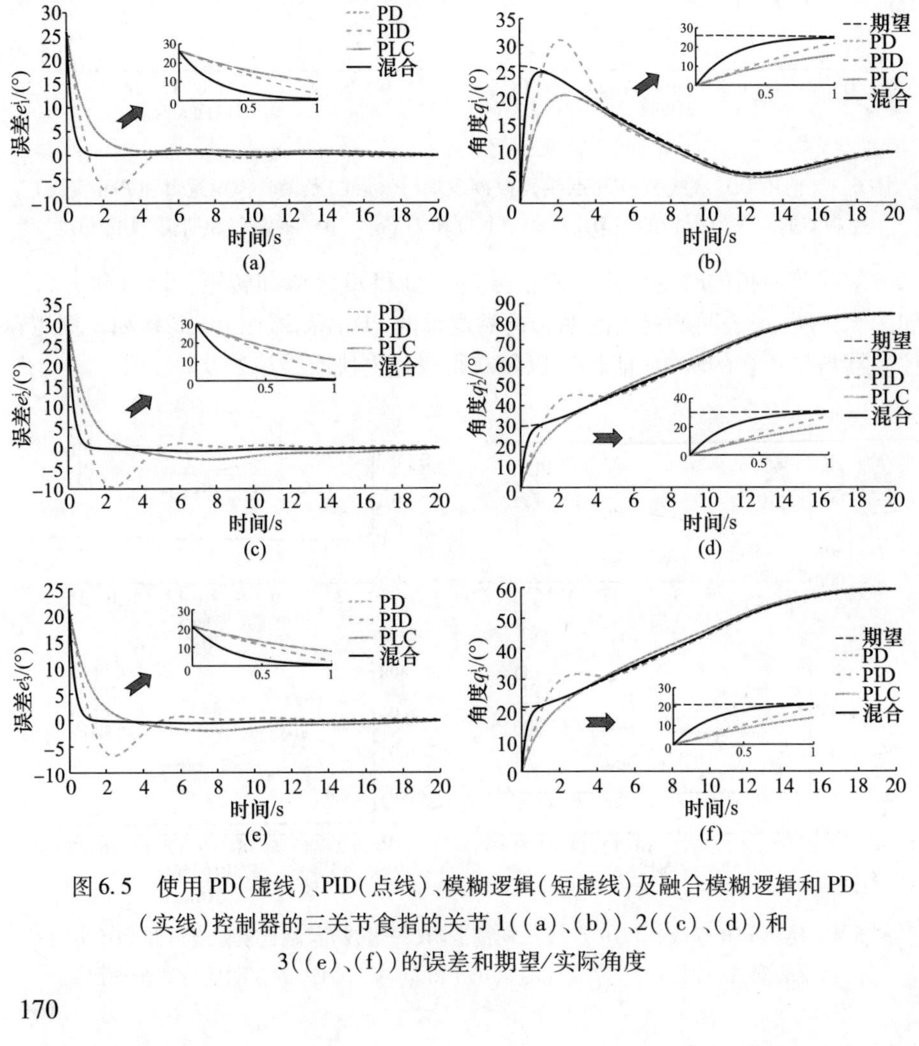

图 6.5　使用 PD(虚线)、PID(点线)、模糊逻辑(短虚线)及融合模糊逻辑和 PD (实线)控制器的三关节食指的关节 1((a)、(b))、2((c)、(d))和 3((e)、(f))的误差和期望/实际角度

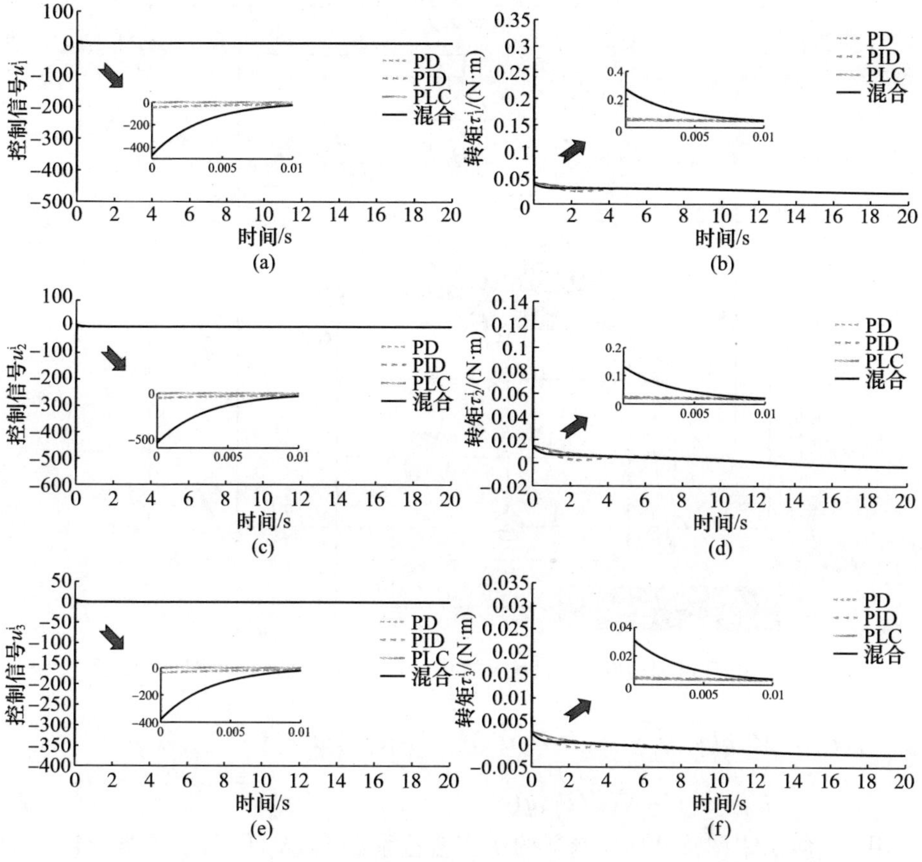

图 6.6　使用 PD(虚线)、PID(点线)、模糊逻辑(短虚线)及融合模糊逻辑和 PD(实线)控制器的三关节食指的关节 1((a)、(b))、2((c)、(d))和 3((e)、(f))的控制信号和转动扭矩

6.2　基于遗传算法的 PID 融合控制策略

图 6.7 是控制信号为

$$\boldsymbol{u}(t) = -\boldsymbol{K}_{\mathrm{P}}\boldsymbol{e}(t) - \boldsymbol{K}_{\mathrm{I}}\int\boldsymbol{e}(t)\,\mathrm{d}t - \boldsymbol{K}_{\mathrm{D}}\dot{\boldsymbol{e}}(t) \tag{6.2.1}$$

的基于融合遗传算法的 PID 控制器框图,具有比例 $\boldsymbol{K}_{\mathrm{P}}(t)$、积分 $\boldsymbol{K}_{\mathrm{I}}(t)$ 和微分 $\boldsymbol{K}_{\mathrm{D}}(t)$ 对角增益矩阵。

重写式(5.1.17)为

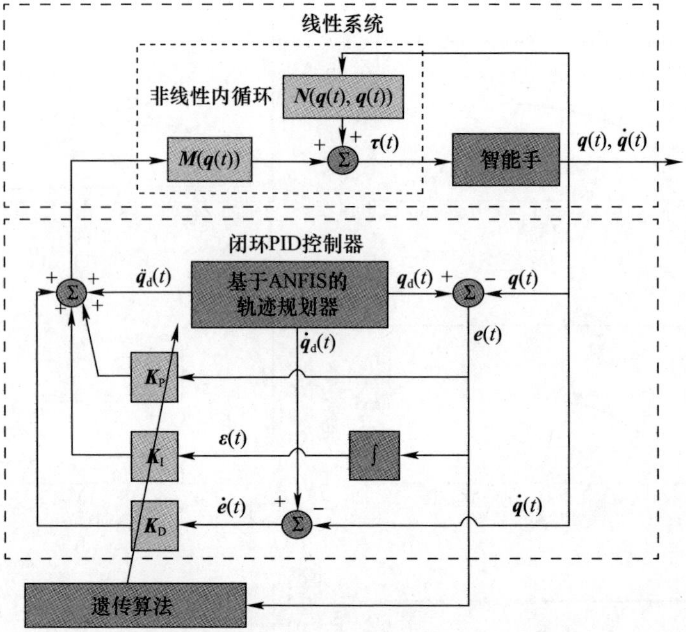

图 6.7　用于 14 自由度五指机械手的基于融合遗传算法的 PID 控制器框图

$$\boldsymbol{\tau}(t) = \boldsymbol{M}(\boldsymbol{q}(t))[\ddot{\boldsymbol{q}}_{\mathrm{d}}(t) + \boldsymbol{K}_{\mathrm{P}}\boldsymbol{e}(t) + \boldsymbol{K}_{\mathrm{I}}\int \boldsymbol{e}(t)\mathrm{d}t + \boldsymbol{K}_{\mathrm{D}}\dot{\boldsymbol{e}}(t)) + \boldsymbol{N}(\boldsymbol{q}(t), \dot{\boldsymbol{q}}(t)] \tag{6.2.2}$$

使用遗传算法来调整 PID 控制器的所有增益系数 K_{P}、K_{I} 和 K_{D}。根据导出的动态和控制模型,在确定参数(K_{P}、K_{I} 和 K_{D})之后,计算扭矩矩阵 $\boldsymbol{\tau}$,然后获得手指 j 的关节 i 的平方跟踪误差 $e_i^j(t)$。因此,总误差 $E(t)$(时间相关函数)定义为

$$E(t) = \int_{t_0}^{t_{\mathrm{f}}} (e_i^j(t))^2 \mathrm{d}t \tag{6.2.3}$$

式中:t_0 和 t_{f} 分别是起始时间和终止时间。如图 4.7 所示,基于之前的研究,获得了调整的对角线参数($\boldsymbol{K}_{\mathrm{P}}(t)$、$\boldsymbol{K}_{\mathrm{I}}(t)$、$\boldsymbol{K}_{\mathrm{D}}(t)$)和通过遗传算法的 PID 控制器的总误差 $E(t)$,并列于表 6.2 中[7]。

表 6.2　遗传算法调谐比例 - 积分 - 微分控制器参数选择及计算总误差

手指	输入			输出
	$\boldsymbol{K}_{\mathrm{P}}$	$\boldsymbol{K}_{\mathrm{I}}$	$\boldsymbol{K}_{\mathrm{D}}$	$E(t)$
实例Ⅰ	[976,956]	[779,279]	[170,236]	0.3107
实例Ⅱ	[988,999]	[78,848]	[80,109]	0.1557

续表

手指	输入			输出
	K_P	K_I	K_D	$E(t)$
实例Ⅲ	[199,198]	[127,157]	[104,102]	0.8100
食指	[794,398,960]	[960,918,914]	[15,59,242]	0.0465
中指	[794,398,960]	[960,918,914]	[15,59,242]	0.1003
无名指	[794,398,960]	[960,918,914]	[15,59,242]	0.0465
小指	[794,398,960]	[960,918,914]	[15,59,242]	0.0607

6.2.1 模拟结果和讨论

为了研究调整的参数范围是否影响总跟踪误差,通过改变双关节拇指的可调参数范围的下限和上限设计了 3 种不同的情况,拇指的 3 种情况——1、2、3 代表 PID 参数 $K_P(t)$、$K_I(t)$ 和 $K_D(t)$ 分别限制在 3 个不同的有界范围[100,1000]、[50,1000]和[100,200]中。图 6.8 和图 6.9 是用于拇指的关节 1 和 2 的 PID 和基于遗传算法的 PID 控制器的跟踪误差和期望/实际角度。这些模拟表明,PID 控制器的参数在大范围[100,1000](情况 1)和[50,1000](情况 2)中比任意选择为 100 有更好的结果。然而,参数在小范围[100,200](情况 3)中比单独的 PID 控制器产生更差的结果。这些结果表明,参数范围越大,总误差越小。情况 1 和 2 说明遗传算法找到一些属于[100,1000]和[50,1000]的参数值来避免局部最小区域。虽然情况 3 的下限值覆盖了 100,但总误差和收敛速度

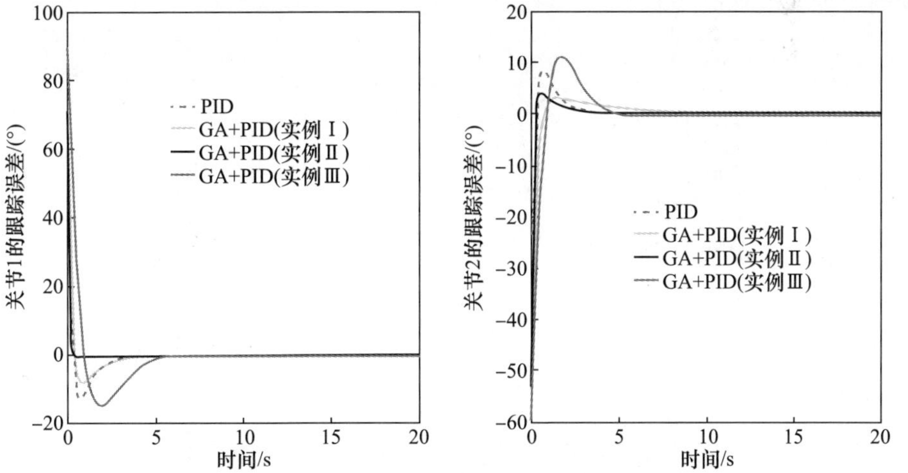

图 6.8 用于拇指的 PID 和基于遗传算法的 PID 控制器关节 1 和 2 的跟踪误差

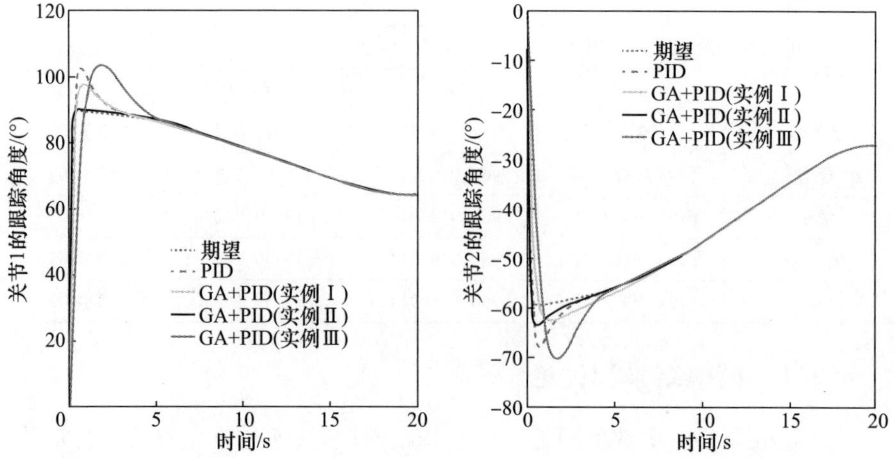

图 6.9 用于拇指的 PID 和基于遗传算法的 PID 控制器关节 1 和 2 的跟踪角度

都比单独的 PID 差,这表明遗传算法在大范围内表现更好,但在边界上搜索效果不佳。为了进一步考虑收敛速度,情况 1 提供了较小的总误差,但与单独的 PID 控制相比,它没有提高收敛速度。然而,情况 2 给出了良好的总误差和收敛速度。情况 3 给出了较差的总误差和收敛速度。总之,这些结果意味着全局最小值可以位于[50,100]和[200,1000]范围内,并且参数范围在遗传算法调整中起重要作用。基于这些发现,对其他的三关节手指使用[50,1000]的参数范围。

图 6.10 至图 6.13 表明了用于其他三关节手指的 PID 和基于遗传算法的 PID 控制器的模拟结果[8]。

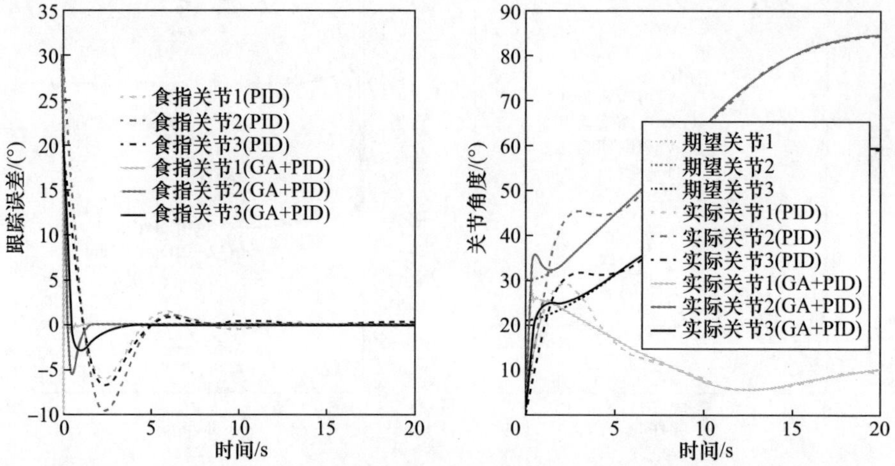

图 6.10 用于食指的 PID 和基于遗传算法的 PID 控制器的跟踪误差和关节角度

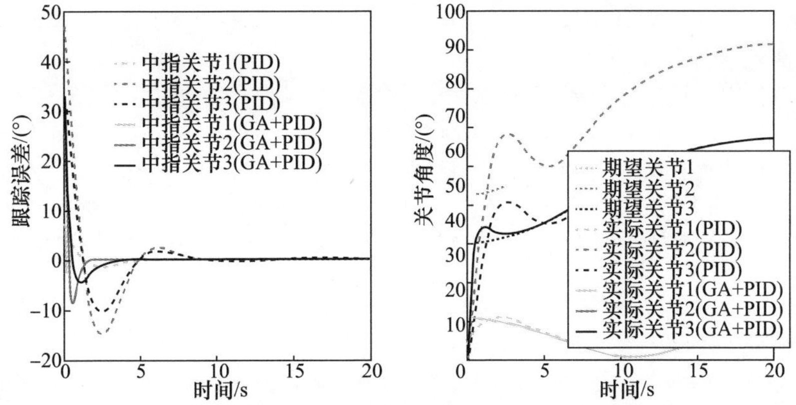

图 6. 11　用于中指的 PID 和基于遗传算法的 PID 控制器的跟踪误差和关节角度

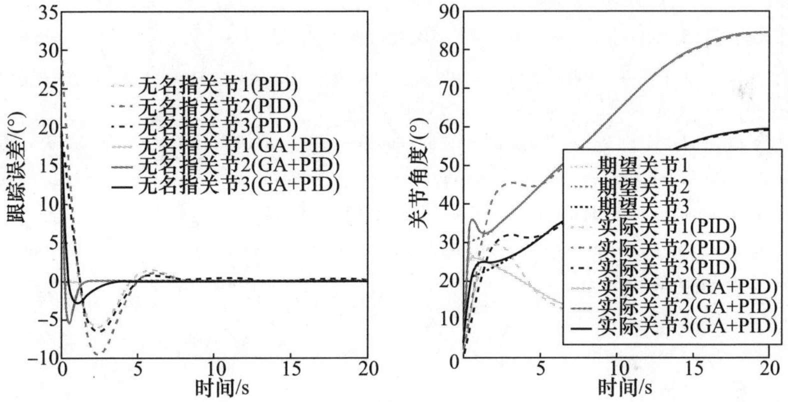

图 6. 12　用于无名指的 PID 和基于遗传算法的 PID 控制器的跟踪误差和关节角度

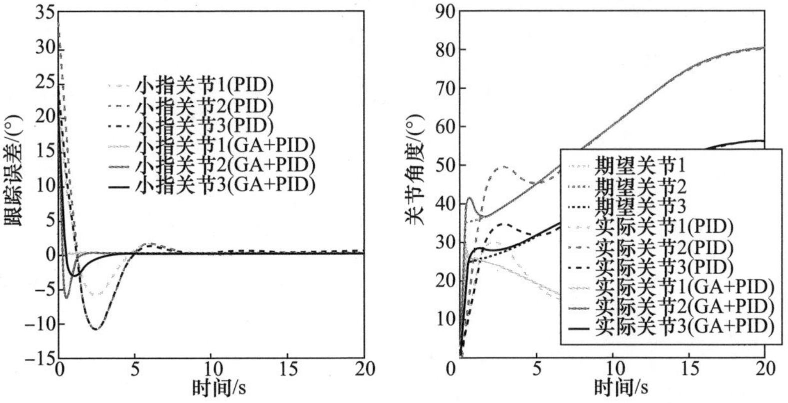

图 6. 13　用于小指的 PID 和基于遗传算法的 PID 控制器的跟踪误差和关节角度

参考文献

[1] E. H. Mamdani and S. Assilian. An experiment in linguistic synthesis with a fuzzy logic controller. *International Journal of Man – Machine Studies*, 7(1):1 – 13, 1975.

[2] S. J. Ovaska, H. F. Van Landingham, and A. Kamiya. Fusion of soft computing and hard computing in industrial applications: An overview. *IEEE Transactions on Systems, Man, and Cybernetics, Part C: Applications and Reviews*, 32(2):72 – 79, May 2002.

[3] S. Arimoto. *Control Theory of Multi – fingered Hands: A Modeling and Analytical – Mechanics Approach for Dexterity and Intelligence*. Springer – Verlag, London, UK, 2008.

[4] P. K. Lavangie and C. C. Norkin. *Joint Structure and Function: A Comprehen – sive Analysis, Third Edition*. F. A. Davis Company, Philadelphia, Pennsylvania, USA, 2001.

[5] C. – H. Chen and D. S. Naidu. "Fusion of fuzzy logic and PD control for a five – fingered smart prosthetic hand," in *Proceedings of the 2011 IEEE International Conference on Fuzzy Systems (FUZZ – IEEE 2011)*, pp. 2108 – 2115, Taipei, Tai – wan, June 27 – 30, 2011.

[6] C. – H. Chen and D. S. Naidu. Hybrid control strategies for a five – finger robotic hand. *Biomedical Signal Processing and Control*, 8(4):382 – 390, July 2013.

[7] C. – H. Chen and D. S. Naidu. "Hybrid genetic algorithm PID control for a five – fingered smart prosthetic hand," in *Proceedings of the Sixth International Conference on Circuits, Systems and Signals (CSS' 11)*, pp. 57 – 63, Vouliagmeni Beach, Athens, Greece, March 7 – 9, 2012.

[8] C. – H. Chen and D. S. Naidu. A modified optimal control strategy for a five – finger robotic hand. *International Journal of Robotics and Automation Technology*, 1(1):3 – 10, November 2014.

第7章　结论与未来展望

7.1　结　　论

本文描述机器假肢手的硬控制策略（如PID、最优控制、自适应控制等）与软控制策略（如自适应神经模糊推理系统（ANFIS）、遗传算法（GA），粒子群优化（PSO）等）的融合。

第2章研究了n维平面转动关节下，双关节拇指和三关节食指的正向运动学、反向运动学和微分运动学模型；使用前向运动学模型推导出了每个指尖（末端执行器）的位置，使用反向运动学从已知的指尖位置（直角坐标空间）推导出了每个手指的关节角（关节空间），进而得到指尖的工作空间；用微分运动学方法计算了指尖的线速度、角速度和加速度。然后利用几何雅可比矩阵求出了每个手指的关节角速度和关节角加速度。此外还推导出了两种轨迹规划函数即三次多项式函数与贝塞尔曲线函数，所得出的结果成功地应用于14自由度五指机械手模型。

在第3章中，使用了一种执行结构的数学模型（该执行结构包含直流电动机和机械齿轮），用拉格朗日方法成功地推导出了两关节和三关节手指的手部运动的动力学方程。第4章中成功开发了软计算（SC）或计算智能（CI）策略，包括模糊逻辑（FL）、神经网络（NN）、ANFIS、禁忌搜索（TS）、遗传算法（GA）、粒子群优化（PSO）算法、自适应粒子群优化（APSO）算法和浓缩型混合优化（CHO）算法，具体如下。

（1）ANFIS和GA：利用ANFIS和GA方法，成功地解决了三关节手指的反向运动学问题。仿真结果表明，遗传算法虽然求出较好的解（误差$\approx 10^{-7}$），但计算时间较长，而ANFIS算法可以求出较好的解（误差$\approx 10^{-4}$），同时耗时较短。因此，本书采用ANFIS方法分析三关节手指的反向运动。

（2）PSO：在PSO动力学研究中，我们认为少量的负向速度是有用的，并取得了较好的结果。也就是说，在这个过程中，大多数粒子都在寻找相同的搜索方向（正权重），但也有少数粒子在寻找不同的搜索空间（负权重）。此外，A－E方法（即在第3章提到的五种PSO方法）均得到了令人满意的结果，但E方法消耗的最多。

(3) APSO：与一般的粒子群算法相比，改进后的粒子群算法改变了更新速度方向，并求解出更好的结果。与一般的粒子群优化算法相比，改变更新速度方向的粒子群优化算法具有更好的稳定性。由于 APSO 方法尝试使用割线平面信息来确定良好的搜索方向，并且对于非常“平滑”的问题，该方法拥有精确定义的极小值，因此求解得到的结果更好。

(4) PSO 在炎症方面应用：结果证明了 PSO 在预测多个选择蛋白 - 配体对综合效应的作用，并使用这些预测来验证我们提出的假说。使用该系统将帮助我们理解多种选择素 - 配体是如何表示和调节体内炎症的。

(5) CHO：仿真结果表明，我们所提出的 CHO 算法融合了 TS 和 PSO 的优点，得到了鲁棒性较好的结果。然而，在更高维问题上，CHO 法依赖于从众多有希望的区域中选择出最有希望的区域，并且选择出的希望区域的常数半径很重要。因此，该参数的选取可以在以后的工作中进行研究。

在第 5 章中，针对前文所提到的机械手，提出了一些硬控制策略，包括了反馈线性化、PD/PI/PID、最优和自适应控制器等。使用 PID 控制器和有限时间内的线性二次型最优控制器，并结合真实数据，得到了 14 自由度五指智能机械手的仿真结果。结果表明在抑制与 PID 控制器相关的超调和瞬态变化时，使用最优控制器可以得到较好的结果。

在第 6 章，研究出了硬控制与软控制的融合策略，该策略充分利用硬控制与软控制中的特征。我们使用真实数据与混合控制策略对具有双关节拇指与三关节食指机械手模型进行了仿真(所采用的混合控制策略包括 PD 控制器与 FL 混合控制、PID 控制器与 GA 算法混合控制等)，结果表明软控制与硬控制一起使用要强于单独使用软控制或硬控制。

7.2 未来发展方向

未来的发展方向可能集中在开发具有 14 和 22 自由度的五指机械手模型和先进的控制策略，来实现触摸、握持、抓取等任务，如图 7.1 所示。

另外，实时控制策略也是未来发展方向之一，具体如下：

(1) 开发一种适用于 14 自由度五指机械手的自适应/鲁棒控制器。在设计自适应/鲁棒控制器时，要考虑实时环境中的不确定性和干扰。

(2) 开发一种适用于 14 自由度五指机械手的 PD/PI 控制器和自适应/鲁棒控制器，设计时要考虑在实时环境下的不确定性因素。

(3) 探索先进的软计算技术，如粒子群优化、遗传算法等，从而提高实时环境下的计算成本。

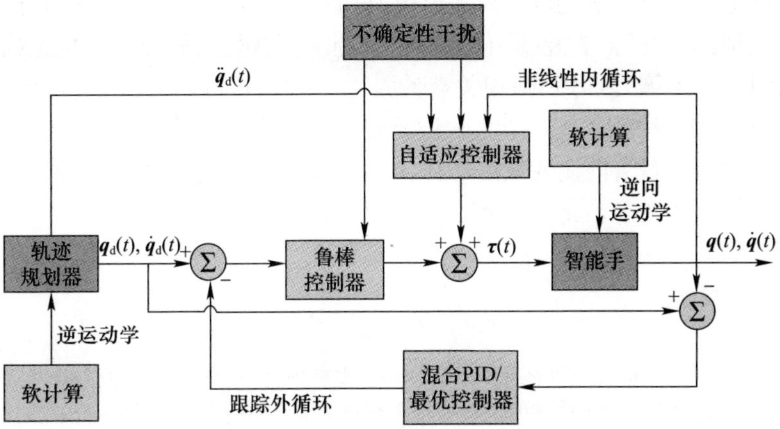

图 7.1　未来混合控制策略框图

（4）开发一种 14 自由度五指手的最优控制器和自适应/鲁棒控制器，设计时要考虑在实时环境下的不确定性因素。

（5）利用遗传算法/粒子群优化算法等软计算方法，开发基于 14 个自由度五指机械手的混合最优控制器和自适应/鲁棒控制器，对考虑实时环境在内的混合控制器参数进行优化。

（6）采用遗传算法/粒子群优化算法等软计算方法，开发基于 14 自由度五手指的混合 PD/PI/PID 控制器和自适应/鲁棒控制器，对考虑实时环境在内的混合控制器参数进行优化。

（7）为 22 自由度五手指机械手的实时使用，优化混合控制策略。

（8）为 22 自由度五手指机械手的实时使用，将集成手部动作、机械设计、抓取操作规划和基于 EMG 的模型以嵌入式的方式分级实现。

在过去的三十年里，为了使机械手实现更多的功能，研究者们开展了很多关于如何使用 EMG 信号的研究。然而，EMG 信号的使用受到类似于人的功能数量的限制，并且使用的电极数量要尽可能的少，最终难以使机械手的外观自然美观。此外，EMG 信号不能向用户提供任何形式的反馈[1]。为了克服表面肌电法的局限性，一种可能的解决方案是神经假体。在这里，我们使用外周神经系统（PNS）和“自然”神经之间的接口，以选择性的方式提取、记录和模拟仿真外周神经系统（PNS）。此外，先进的生物相容性神经接口可以刺激传入神经，使假肢运动控制向“自然”肌电控制过渡，为使用者提供一些感觉反馈。基于上述分析，两种可能控制机械手的方式是基于 EMG 的简单非侵入式的控制和更加复杂的植入式控制。

根据美国国家科学院凯克未来倡议,正如文献[2]中所述“智能假肢是身体和大脑的辅助设备”。它还指出“我们可以制造更智能的假肢……集工程、医学和社会科学于一体”。另一则相关新闻是关于大脑控制的机械手(价值1.2亿美元),其代表了约翰霍普金斯大学神经假肢技术的进展[3]。

其他有趣或者有前途的想法是在3D打印,如从文献[4]中看到的“3D打印的假肢帮助儿童运动员比赛”。

参考文献

[1] M. Zecca, S. Micera, M. C. Carrozza, and P. Dario. Control of multifunctional prosthetic hands by processing the electromyographic signal. *Critical Reviews in Biomedical Engineering*, 30:459 – 485, 2002 (Review article with 96 references).

[2] Arnold and Mabel Beckman Center of the National Academies. *NAKFI: Smart Prosthetics: Exploring Assistive Devices for Body and Mind: Task Group Summaries*, Irvine, California, USA, November 9 – 11, 2007.

[3] S. Grobart. This guy has a thought controlled robotic arm, November 2015.

[4] F. Imbert. 3 – D printed prosthetics help kid athletes compete, November 2015.

结　束　语

本书的参考依据：

以下是作者及其同事的出版物（按时间顺序排列，未列出提交给资助机构的大量季度报告），构成了该书的主要基础：

1. J. C. K. Lai, M. P. Schoen, A. Perez – Gracia, D. S. Naidu, and S. W. Leung, "Prosthetic devices: Challenges and implications of robotic implants and biological interfaces," Spe – cial Issue on Micro and Nano Technologies in Medicine, Proceedings of the Institute of Mechanical Engineers (IMechE, London, UK, Part H, Jounal of Engineering in Medicine, 221(2): 173 – 183, 2007. Listed as 1 of 20 in *Top 20 Articles, in the Domain of Article 17385571, Since its Publication (2007) according to BioMedLib: "Who is Publishing in My Domain ?"* as on September 22, 2014 and as 1 of 20 as on March 17, 2015.

2. C. – H. Chen, K. W. Bosworth, and M. P. Schoen, "Investigation of particle swarm optimization dynamics," in *Proceedings of International Mechanical Engineering Congress and Exposition (IMECE) 2007*, Seattle, Washington, USA, November 11 – 15, 2007 (No. IMECE2007 – 41343).

3. C. – H. Chen, K. W. Bosworth, M. P. Schoen, S. E. Bearden, D. S. Naidu, and A. PerezGracia, "A study of particle swarm optimization on leukocyte adhesion molecules and control strategies for smart prosthetic hand," in *2008 IEEE Swarm Intelligence Symposium (IEEE SIS08)*, St. Louis, Missouri, USA, September 21 – 23, 2008.

4. D. S. Naidu, C. – H. Chen, A. Perez – Gracia, and M. P. Schoen, "Control strategies for smart prosthetic hand technology: An overview," in *Proceedings of the 30th Annual International IEEE EMBS Conference, Vancouver, Canada, pp. 4314 – 4317, August 20 – 24, 2008 (In Top 20 Articles, in the Domain of Article 19163667, Since its Publication (2008) according to BioMedLib: "Who is Publishing in My Domain?,"* Ranked as No. 8 of 20 as on August 1, 2014, Ranked as No. 9 of 20 as on May 4, 2015).

5. C. – H. Chen, D. S. Naidu, A. Perez – Gracia, and M. P. Schoen, "Fusion of hard

and soft control techniques for prosthetic hand," *in Proceedings of the International Association of Science and Technology for Development (IASTED) International Conference on Intelligent Systems and Control (ISC 2008)*, Orlando, Florida, USA, pp. 120 – 125, November 16 – 18, 2008.

6. C. – H. Chen, K. W. Bosworth, and M. P. Schoen, "An adaptive particle swarm method to multiple dimensional problems," *in Proceedings of the International Association of Science and Technology for Development (IASTED) International Symposium on Computational Biology and Bio informatics (CBB 2008)*, Orlando, Florida, USA, pp. 260 – 265, November 16 – 18, 2008.
7. C. – H. Chen and D. S. Naidu, *Intelligent Control for Smart Prosthetic Hand Technology – Phase 1 – Year 1 – Annual*, Year 1: Annual Report, Measurement and Control Engineering Research Center (MCERC), College of Engineering, Idaho State University, Pocatello, Idaho, USA, August 5, 2008.
8. C. – H. Chen, "Hybrid control strategies for smart prosthetic hand," PhD Dissertation, Idaho State University, Pocatello, Idaho, USA, May 2009.
9. C. – H. Chen, D. S. Naidu, A. Perez – Gracia, and M. P. Schoen, "Hybrid control strategy for five – fingered smart prosthetic hand," in *The 48th IEEE Conference on Decision and Control (CDC) and 28th Chinese Control Conference (CCC)*, Shanghai, P. R. China, pp. 5102 – 5107, December 16 – 18, 2009.
10. C. – H. Chen, M. P. Schoen, and K. W. Bosworth, "A condensed hybrid optimization algorithm using enhanced continuous tabu search and particle swarm optimization," in *Proceedings of ASME 2009 Dynamic Systems and Control Conference (DSCC)*, Holly – wood, California, USA, October 12 – 14, 2009 (No. DSCC2009 – 2526).
11. C. – H. Chen, D. S. Naidu, A. Perez – Gracia, and M. P. Schoen, "Hybrid optimal control Strategy for smart prosthetic hand," in *Proceedings of ASME 2009 Dynamic Systems and Control Conference (DSCC)*, Hollywood, California, USA, October 12 – 14, 2009 (No. DSCC2009 – 2507).
12. C. – H. Chen, D. S. Naidu, A. Perez – Gracia, and M. P. Schoen, "Hybrid adaptive control strategy for smart prosthetic hand," in *The 31st Annual International Conference of the IEEE Engineering Medicine and Biology Society (EMBC)*, Minneapolis, Minnesota, USA, pp. 5056 – 5059, September 2 – 6, 2009.
13. C. – H. Chen, D. S. Naidu, A. Perez – Gracia, and M. P. Schoen, "Hybrid of hard control and soft computing for five – fingered prosthetic hand," in *Graduate Student Research and Creative Excellence Symposium, Idaho State University*, Poca-

tello, Idaho, USA, April 10, 2009 (Poster presentation).
14. C. – H. Chen, K. W. Bosworth, and M. P. Schoen, "An investigation of particle swarm optimization dynamics," in *Graduate Student Research and Creative Excellence Symposium, Idaho State University*, Pocatello, Idaho, USA, April 10, 2009 (Poster presentation).
15. C. – H. Chen and D. S. Naidu, *Intelligent Control for Smart Prosthetic Hand Technology – Phase 1 – Year2 – Annual*, Annual Research Report, Measurement and Control Engineering Research Center (MCERC), College of Engineering, Idaho State University, Pocatello, Idaho, USA, August 15, 2009.
16. C. – H. Chen and D. S. Naidu, *Intelligent Control for Smart Prosthetic Hand Technology – Phase 1 – Final*, Final Research Report, Measurement and Control Engineering Research Center (MCERC), College of Engineering, Idaho State University, Pocatello, Idaho, USA, August 22, 2009.
17. C. – H. Chen, D. S. Naidu, and M. P. Schoen, "An adaptive control strategy for a five – fingered prosthetic hand," in *The 14th World Scientific and Engineering Academy and Society (WSEAS) International Conference on Systems, Latest Trends on Systems (Volume II), Corfu Island, Greece*, pp. 405 – 410, July 22 – 24, 2010.
18. C. – H. Chen and D. S. Naidu, "Optimal control strategy for two – fingered smart prosthetic hand," in *Proceedings of the International Association of Science and Technology for Development (IASTED) International Conference on Robotics and Applications (RA2010)*, Cambridge, Massachusetts, USA, pp. 190 – 196, November 1 – 3, 2010.
19. C. – H. Chen, D. S. Naidu, and M. P. Schoen, "Adaptive control for a five – fingered prosthetic hand with unknown mass and inertia," *World Scientific and Engineering Academy and Society (WSEAS) Journal on Systems*, 10: 148 – 161, May 2011.
20. C. – H. Chen and D. S. Naidu, "Fusion of fuzzy logic and PD control for a five – fingered smart prosthetic hand," in *Proceedings of the 2011 IEEE International Conference on Fuzzy Systems (FUZZ – IEEE 2011)*, Taipei, Taiwan, pp. 2108 – 2115, June 27 – 30, 2011.
21. D. S. Naidu and C. – H. Chen, "Automatic control techniques for smart prosthetic hand technology: An overview," chapter 12 in *Distributed Diagnosis and Home Healthcare (D2H2), Vol. 2*, edited by U. R. Acharya, F. Molinari, T. Tamura, D. S. Naidu, and J. Suri, American Scientific Publishers, Stevenson Ranch, Cali-

fornia, USA, pp. 201 – 223, 2011.

22. C. – H. Chen and D. S. Naidu, *Intelligent Control for Smart Prosthetic Hand Technology – Phase 2 – Year1 – Annual*, Annual Research Report, Measurement and Control Engineering Research Center (MCERC), College of Engineering, Idaho State University, Pocatello, Idaho, USA, April 25, 2011.
23. C. – H. Chen and D. S. Naidu, "Hybrid genetic algorithm PID control for a five – fingered smart prosthetic hand;" in *Proceedings of the 6th International Conference on Circuits, Systems and Signals (CSS' 11)*, Vouliagmeni Beach, Athens, Greece, pp. 57 – 63, March 7 – 9, 2012.
24. C. – H. Chen and D. S. Naidu, *Intelligent Control for Smart Prosthetic Hand Technology – Phase 2 – Year2 – Annual*, Annual Research Report, Measurement and Control Engineering Research Center (MCERC), College of Engineering, Idaho State University, Pocatello, Idaho, USA, April 25, 2012.
25. C. – H. Chen and D. S. Naidu, *Intelligent Control for Smart Prosthetic Hand Technology – Phase 2 – Year3 – Annual*, Annual Research Report, Measurement and Control Engineering Research Center (MCERC), College of Engineering, Idaho State University, Pocatello, Idaho, USA, December 15, 2012.
26. C. – H. Chen and D. S. Naidu, *Intelligent Control for Smart Prosthetic Hand Technology – Phase 2 – Final*, Final Research Report, Measurement and Control Engineering Research Center (MCERC), College of Engineering, Idaho State University, Pocatello, Idaho, USA, December 19, 2012.
27. C. – H. Chen, "Homocysteine – & connexin 43 – regulated mechanisms for endothelial wound healing;" PhD Dissertation, Idaho State University, Pocatello, Idaho, USA, May 2013.
28. C. – H. Chen and D. S. Naidu, "Hybrid control strategies for a five – finger robotic hand," *Biomedical Signal Processing and Control*, 8:382 – 390, July 2013.
29. C. Potluri, M. Anugolu, M. P. Schoen, D. S. Naidu, A. Urfer, and S. Chiu, "Hybrid fusion of linear, non – linear and spectral models for the dynamic modeling of sEMG and skeletal muscle force: An application to upper extremity amputation," *Computers in Biology and Medicine: An International Journal*, 43(11): 1815 – 1826, November 2013.
30. C. – H. Chen and D. S. Naidu, "A modified optimal control strategy for a five – fingerrobotic hand," *International Journal of Robotics and Automation Technology*, 1:3 – 10, November 2014.

致　谢

这项研究由美国陆军部资助，编号为 W81XWH－10－1－0128，并由美国陆军医学研究与采购部门管理。该部门位于美国马里兰州德特里克堡钱德勒街820号，邮政编码是21702－5014。这些信息不一定能反映出政府的立场或政策，也不代表该研究得到了官方认可。就本书而言，信息来源于发表的新闻稿、文章、手稿、小册子、广告、静态/动态图片、演讲、行业协会的会议记录等。

作 者 简 介

Cheng－Hung Chen 获得台湾嘉义“国立中正大学”(CCU)机械工程学士学位,台湾新竹“国立清华大学”动力机械工程硕士学位(固体力学专业)。陈博士来到美国,开始在马萨诸塞州剑桥市的哈佛大学学习生物学和化学。然后,他从波士顿驱车 2550 英里前往波卡特洛,在爱达荷州立大学(ISU)大学攻读双博士学位(工程和应用科学与生物科学)、工商管理硕士学位(市场营销和管理专业)和化学学士学位,并辅修心理学。随后,陈博士作为马萨诸塞大学(UMASS)阿默斯特分校的博士后研究助理,开发了一款适用于血液透析患者的控制器。

Cheng－Hung Chen 在慕尼黑工业大学(TUM)参加了地球空间科学和技术的硕士课程,并在德国慕尼黑飞行学院参加了私人飞行员执照(PPL)的培训。他扩展了自己在航空航天领域的知识面,包括先进轨道力学、卫星导航、轨道动力学和机器人技术、航天器技术、摄影测量、(微波)遥感技术、图像处理、信号处理和地球科学等。目前,他在 Synova Inc. 的美国微加工中心(MMC)担任应用工程师,该公司是一家总部位于瑞士 Duillier 的 Laser MicroJet© 技术公司。他在做激光微加工测试,设计最佳参数集,优化激光工艺,以最大限度地提高性能、质量和精度,来满足航空航天、医疗保健、钻石和珠宝、能源、工具制造和半导体行业的客户需求。

陈博士是 IEEE 高级会员,他在工程学上的研究方向包括固体力学、控制系统和航空航天、有限元建模、数值建模、复合材料力学、静态/动态分析、断裂力学、振动分析、热传导分析、结构测试和自动控制、PID 控制、最优控制、自适应/鲁棒控制、模糊逻辑、神经网络、禁忌搜索、遗传算法、粒子群优化、机器人技术、航天器技术和轨道力学等。他在生物学上的研究兴趣是人体生理学,包括心血管疾病、神经认知疾病和癌症、微循环和血液透析等。他在工程和生物学领域的国际出版物中发表了 33 篇文章,并担任了 10 种生物医学工程期刊的编委/审稿人。

他的职业生涯的总体目标是:①成为一家将学术、工业、政治、市场和教育结合在一起的航空航天非盈利基金会的首席执行官;②从太空探索和收获新的珍贵材料;③开发微重力下的航空航天医学(太空医院、太空救护车、太空工厂、太空运输等);④在外太空开展人类活动。

Desineni Subbaram Naidu 在印度安得拉邦蒂鲁伯蒂的斯里文卡特斯瓦拉大学获得了电子工程学士学位，并在印度卡拉格普尔的印度理工学院（IIT）获得了电气工程（控制系统工程）硕士学位和电气工程博士学位。

目前，Naidu 教授是美国明尼苏达大学德卢斯分校（Minnesota Duluth）的由明尼苏达电力公司 Jack F. Rowe 捐赠的基金会主席和电气工程教授。以下为其工作过的地点：印度理工学院、弗吉尼亚州汉普顿市 NASA 兰利研究中心制导与控制处、弗吉尼亚州诺福克市老道明大学、位于俄亥俄州 WPAFB 的美国空军研究实验室先进飞行研究中心、挪威特隆赫姆挪威科技大学船舶和海洋结构中心、苏黎世瑞士联邦理工学院测量与控制实验室、中国的南通大学、珀斯的西澳大利亚大学、澳大利亚的阿德莱德的南澳大利亚大学工业和应用数学中心、中国上海的华东师范大学应用与跨学科数学研究中心。Naidu 教授曾在爱达荷州立大学任电气工程教授、工程学院副院长、工程学院院长与测量与控制工程研究中心主任。

Naidu 教授的主要教学和研究领域包括电气工程、电网、线性控制系统、数字控制系统、最优控制系统、非线性控制系统、智能控制系统，生物医学和工程，包括假肢、机器人、机电一体化、制导和控制、航空航天系统、轨迹优化、火星任务中的飞行制动、轨道力学、控制理论和应用的奇异扰动和时间尺度（SPaTS），金属电弧焊（GMAW）中气体的传感与控制。

Naidu 教授获得了多个奖项和个人荣誉，其中包括两次美国国家科学院国家研究委员会的高级研究助理，当选为电气与电子工程师学会（IEEE）院士，当选为世界创新基金会院士，以及爱达荷州立大学（ISU）杰出研究员和杰出研究员。他曾在期刊和会议出版物中发表过 200 多篇文章，并出版过 8 本书。他曾担任过几种期刊的编委员会成员，包括 IEEE 学报的《自动控制与最优控制：应用与方法》（Wiley）。